CRTD-Vol. 37

An ASME Research Report

SCREW-THREAD GAGING SYSTEMS FOR DETERMINING CONFORMANCE TO THREAD STANDARDS

PREPARED BY
Steering Committee on "Screw-Thread Gaging Systems"

Leroy S. "Skip" Fletcher, Chair
Texas A&M University

Bruce W. Christ
Institute for Better Construction Materials

Kenneth E. McCullough
Consultant

Thomas Middleton
McDonnel Douglas Corporation

FOR THE
ASME Center for Research and Technology Development

THE AMERICAN SOCIETY OF MECHANICAL ENGINEERS
Three Park Avenue / New York, N.Y. 10016

NOTICE

This report was prepared as an account of work coordinated through the American Society of Mechanical Engineers' (hereafter the "Society") Center for Research and Technology Development in the course of performing work contracted for and sponsored by the National Institute of Standards and Technology (hereafter the "Sponsor"). The opinions expressed in this report do not necessarily reflect those of the Sponsor, the Society, or others involved in the preparation or review of this report, or any agency thereof, and reference to any specific product, service, process, or method does not constitute an implied or expressed recommendation or endorsement of it. Further, the Sponsor and the Society make no warranties or representations, expressed or implied, as to the fitness for particular purpose or merchantability of any product, apparatus, or service, or the usefulness, completeness, or accuracy of any process, methods or other information contained, described, disclosed, or referred to in this report. The Sponsor and the Society make no representation that the use of any product, apparatus, process, method, or other information will not infringe privately owned rights and will assume no liability for any loss, injury, or damage resulting from, or occurring in connection with, the use of information contained, described, disclosed, or referred to in this report.

ISBN No. 0-7918-1228-6

SCREW-THREAD GAGING SYSTEMS FOR DETERMINING CONFORMANCE TO THREAD STANDARDS

EXECUTIVE SUMMARY

Screw-thread gaging systems which appear in consensus American National Standards and in non-consensus Federal Standards for procurement by the federal government have been examined. The thread standards define the threads as tolerance zones, define the gage configuration and use and establish acceptability criteria for the product thread with gaging systems. The gaging systems in each type of standard are essentially the same. Both the consensus and non-consensus standards for screw-thread gaging systems are incorporated by reference into the non-consensus military specifications. The three gaging systems examined — System 21, System 22, and System 23 — are used for inspecting unified inch and metric screw-threads. Examination of the gages specified for System 21 inspection verifies a technical point of long standing, namely, that inspection by System 21 may not assure conformance to dimensional specifications in thread standards for the thread element, minimum external pitch diameter or maximum internal pitch diameter. The reason is that the gages specified for System 21 inspection are not designed to inspect for the element pitch diameter. Systems 22 and 23 are more severe inspections than System 21, and both Systems 22 and 23 specify gages which can successfully inspect for the thread element, minimum external pitch diameter or maximum internal pitch diameter but may not inspect for the minimum material tolerance zone. All three gaging systems can assure conformance to dimensional specifications for maximum pitch diameter over the gage length inspected, which is usually less than the entire product thread length. Size inspections for all thread elements and thread characteristics specified in System 23 can assure conformance to dimensional specifications but maybe not the tolerance zone. When properly used, all three screw thread gaging systems are satisfactory for their intended use.

There has been an increase in the use of indicating gages for inspections by all three screw-thread gaging systems. The driving forces for this transition, resulting from the increased demand of aerospace and automotive sectors of the economy, has been driven by quality assurance featuring the use of statistical process control to assure thread acceptability in the manufacturing operations.

Recommendations of this study include improvement in the clarity of existing standards for screw-thread gaging systems, e.g., nomenclature; acceptability in comparison to dimensional specifications; meaning of inspection results; differences among inspection results from Systems 21, 22, and 23, improved education about screw-thread gaging systems for technologists who design/procure/manufacture screw-thread products, attention to some metrology concerns related to design of segments and rolls of indicating gages, development of a data base on how variations in thread dimensions affect mechanical performance of threaded assemblies, and promotion of research in support of developing laser and holographic screw-thread gaging systems.

Independent Peer Reviewers

Mr. Bengt Blendulf, Acting Director, Office of Continuing Education, Clemson University

Mr. Harry S. Gibson, Group Engineer, Standards and Specifications Engineering, Lockheed Martin Aeronautical Systems

Mr. William Matievich, Director of Technology, SPS Technologies

Mr. Charles J. Wilson, Director of Engineering, Industrial Fasteners Institute

CONTENTS

LIST OF FIGURES

LIST OF TABLES

I. INTRODUCTION

A. Organization of Study of Screw-Thread Gaging Systems

1. Study Initiation

The National Institute of Standards and Technology (NIST) initiated this study of screw-thread gaging systems for threaded products in order to address issues set forth by the community which generates, maintains, and uses standards for screw-thread technology. The American Society of Mechanical Engineers Center for Research and Technology's (ASME/CRTD) role as administrator of this study was formalized by letter dated May 13, 1994 from the National Institute of Standards and Technology.

2. ASME/CRTD Panel to Identify/Investigate Issues

ASME/CRTD convened a panel which instituted several administrative procedures to identify and investigate issues about threaded products linked to inspection standards and specifications, metrology, and mechanical performance.

A three-person Steering Committee with expertise in standardization documents and screw-thread technology was formed to oversee the study.

A Principal Investigator was appointed to gather technical data by telephone, face-to-face interviews, and correspondence, and to prepare reports of findings.

A meeting of interested parties was convened on August 12, 1994 in which all parties were invited to: 1) identify issues linked to screw-thread inspection technology, 2) provide technical documentation about the issues, and 3) suggest options for resolving the issues. A form was provided to interested parties to complete and submit to ASME/CRTD.

NIST and ASME/CRTD issued a joint press release about this study in the fall of 1994. The press release invited interested parties to contact the ASME/CRTD office in Washington, DC or ASME/CRTD's Principal Investigator. Interested parties were invited to: 1) identify issues linked to screw-thread inspection technology, 2) submit technical documentation about issues, and 3) suggest options for resolving the issues.

B. Scope and Objectives of Study

1. Scope

The national community which generates, maintains, and uses standards for screw-thread technology has been involved since about 1977 in controversy over certain gaging systems in certain inspection standards. Mechanical performance of assemblies made with inspected threaded products has been the focus of allegations about non-conforming threaded products failing and causing incidents. This study intends to provide an objective identification of the issues, to conduct an open investigation for technical data which address the issues, and to provide options for resolving the issues. This study was accomplished outside of the normal standards consensus process.

This study examines: 1) inspection standards for unified inch screw-threads and related standardization documents, 2) screw-thread inspection methods/gaging systems in inspection standards, 3) certain gages and measuring equipment used for inspection of threaded products, and 4) technical data about the mechanical performance of assemblies, as the data relate to the selected inspection method. This study does not focus on plated or coated threads, or metric screw-threads, although the United States inspection standards for metric screw-threads are similar to those for unified inch screw-threads.

2. Objectives

a. Examine standards and metrology issues as they relate to threaded products.

b. Set forth unresolved issues and make recommendations for consideration.

c. Clarify what is meant by an *acceptability standard* for screw thread inspection.

d. Insofar as possible, determine the susceptibility to failure for threaded products which satisfy inspection criteria, as defined in the selected screw-thread gaging system.

e. Identify and investigate issues linked to:

(1) inspection standards and inspection methods/gaging systems for threaded products

(2) gages and measuring equipment used for inspection of threaded products

(3) the concepts of *dimensional conformance* and *acceptability*

(4) the mechanical performance of threaded assemblies.

C. Overview of Threaded Products

1. Importance of Consistency in Threaded-Product Standards

A huge volume of screw-thread products is used throughout the industrial/technological world. For example, a facility supplying assembled air conditioners to an automotive assembly plant might have a daily throughput of one hundred thousand threaded products. Threaded products literally hold the industrial/technological world together.

Screw-thread technology is complex. This complexity arises in part from the basic design of a screw thread, which involves a helical geometry[1]. In the manufacturing of threaded products, the influence of

[1]The definition of a screw thread is found in ANSI/ASME B1.7M-1984 — *NOMENCLATURE, DEFINITIONS, AND LETTER SYMBOLS FOR SCREW THREADS* [1][*]: *screw thread* — a continuous and projecting helical ridge usually of uniform section on a cylindrical or conical surface.

service wear on dimensional accuracy of tools adds to the complexity. Another problem arises from the influence of wear/rust/galling on the accuracy of inspection gages and measuring equipment. The need for critical assembly, e.g., appropriate preload or torque in applications demanding a high level of mechanical reliability and durability, is yet another facet of the complexity of thread gaging or measuring.

Figure I-1 shows the basic components of a screw thread assembly, a product with an internal screw thread and a product with an external screw thread, and some basic screw-thread notation. (The symbols are not all current, but are defined in the Figure.)

A sizable national and trans-national community generates, maintains, and uses standardization documents for screw-thread technology. Custodians of standardization documents for screw-thread technology conduct ongoing activities intended to promote consistency among their own inter-related documents, as well as to promote consistency between their documents and the standardization documents handled by other custodians. Inter-custodian activities sometimes occur which result in certain standardization documents managed by one custodian becoming incorporated by reference into some of the documents handled by other custodians.

Consistency in this network of standardization documents for threaded products expedites the successful management of large and complex engineering projects. Such projects usually involve numerous applications of various kinds of threaded products. Consistency in standardization documents assures: 1) unified thread forms and widely-recognized thread series designations, 2) agreed-upon nomenclature and definitions for the features of threaded products, 3) controlled variation of thread elements and thread characteristics; and 4) assembly of externally-threaded products from one manufacturer with internally-threaded products from another manufacturer. Interestingly, the standards/specifications for inspection of screw thread products address only dimensional conformance.

2. Kinds of Threaded Products

Threaded products are diverse in size, shape, and thread form. Familiar threaded products are nuts and bolts. Threaded products may also be components in structures, such as cylindrical flanges with internally-threaded holes used in tubular systems for scaffolds. Threaded couplings are used in hydraulic systems and some types of pipelines.

This study focuses on ***unified inch screw threads***, and specifically on UN, UNR, and UNJ thread forms. The basic profile for UN and UNR 60° screw threads appears in Figure I-2, which is taken from ASME B1.1 1989 — *Unified Inch Screw Threads (UN and UNR Thread Form)* [3]. Since 1949, this standard for unified inch screw-threads has facilitated general interchangeability of threads through the standardization of thread form, diameter-pitch combinations, and limits of size.

3. Applications of Threaded Products

The application of a screw-thread product usually determines the *thread series* and the *thread class*.

---

[*] Numbers in square brackets designate references at the end of the report.

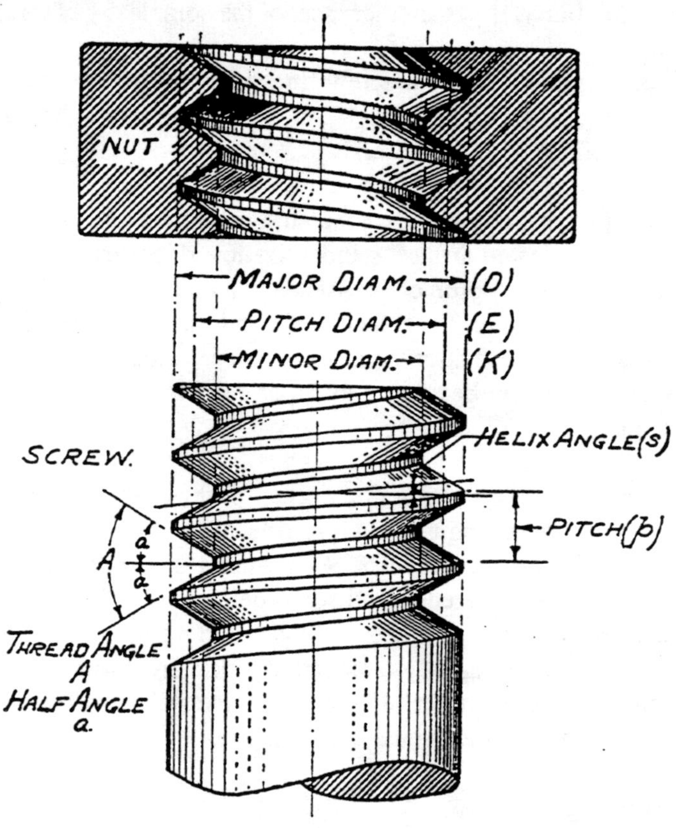

For use in formulas for expressing relations of screw threads, and for use on drawings and for similar purposes, the following symbols should be used:

Major diameter	D
Corresponding radius	d
Pitch diameter	E
Corresponding radius	e
Minor diameter	K
Corresponding radius	k
Angle of thread	A
One half angle of thread	a
Number of turns per inch	N
Number of threads per inch	n
Lead	$L=\frac{1}{N}$

I-1. Basic components of a screw-thread assembly, and screw-thread notation. [2] For current symbols, see ANSI B1.7M [1]. (Helix angles are now called lead angles.)

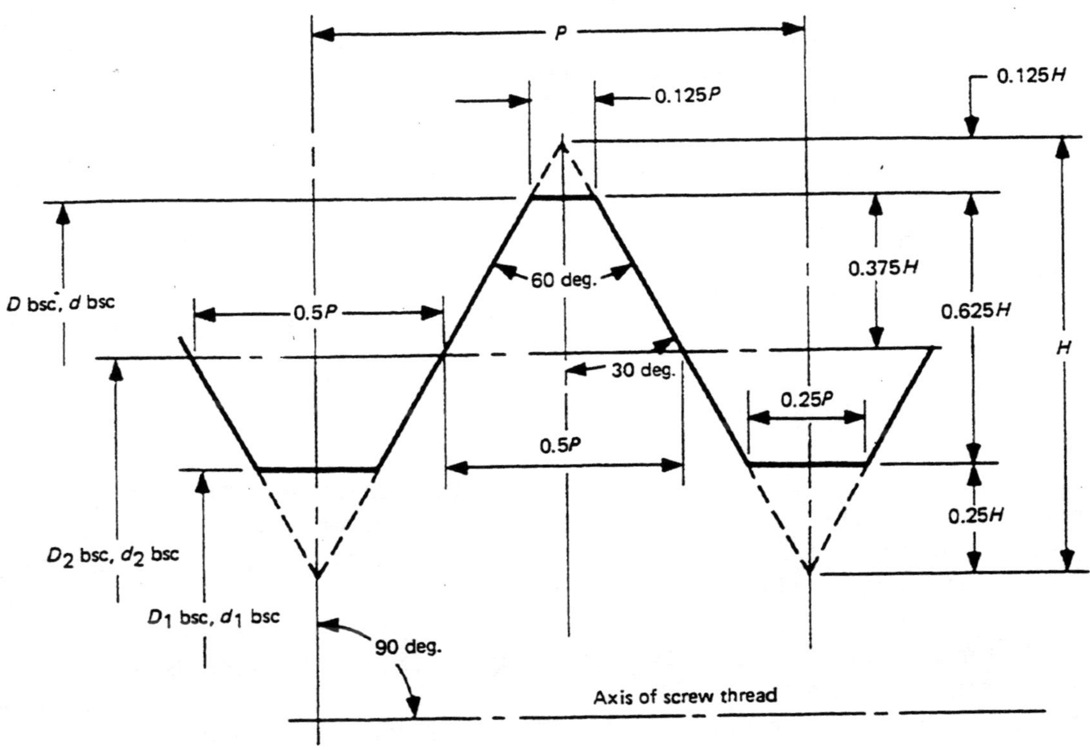

I-2. Basic profile for UN, UNR, and metric screw-threads [3].

Thread series are groups of diameter-pitch combinations distinguished from each other by the number of threads per inch applied to a series of specific diameters. A thread series is designated with the notation such as: **¼-20UNC-2A** or **0.250-28UNJF-3A**. There are two general series classifications: standard and special. Thread series are described in ANSI/ASME B1.7M-1984 (Table 2 on pages 15-16) [1], and in ASME B1.1-1989 (on pages 3-6) [3].

Thread classes are distinguished from one another by the tolerance and allowance, as shown in Figure I-3. Thread classes of mating parts determine fit. Based on applications and design criteria, threads are selected from Class 3A and Class 3B/fine series threads, Class 2A and Class 2B/fine or coarse series threads, and Class 1A and Class 1B/coarse series threads. Thread classes are discussed in ASME B1.1-1989 (pages 6-7) [3]. (In this standard, "A" refers to an external thread; "B" to an internal thread.) Designers/procurers of threaded products make decisions about the selection of *thread series* and *thread class*.

> NOTE Military applications of threaded products are classified in MIL-S-8879C-1991 [4] in two "Application Categories": 1) *Safety Critical Thread*, or 2) *Other Thread*. The basis for determining the application category of threads is set forth in MIL-S-8879-C-1991, which requires "Durability and Damage Tolerance Analyses" and "Failure Modes Effects and Criticality Analyses", as well as "Critical Parts Identification". Many military applications of threaded products require the equivalent of a Class 3A/3B fit.

4. Different Standards Used

A vast network of standardization documents is the foundation of screw-thread technology. A complete set of interrelated standards for the design, manufacture, and inspection of screw-thread products includes at least the following four types of standards: 1) a screw thread standard, 2) a nomenclature standard, 3) an inspection standard, and 4) a gage standard[2].

At least four major sources of screw-thread technology standardization documents exist for some or all of the four categories of standards cited above: A) American National Standards, B) Federal Procurement Standards, C) Military Specifications, and D) ISO Standards. Provisional National Aerospace Standards (see footnote, Section II-C of this report) were prepared in 1992, but not issued.

American National Standards are *consensus* standards and ISO standards are *non-consensus* standards. Federal Procurement Standards and Military Specifications are *non-consensus* standardization documents, and the provisional National Aerospace Standards are also *non-consensus* documents.

Understanding the history of domestic screw-thread standardization documents provides informative perspective on the issues being addressed by this study. Table I-1 gives an overview of the historical development of the three sets of domestic screw thread standardization documents used in the United

[2]Correct terminology is critical for clarity in discussions of screw-thread technology. The terminology used throughout this report is found in ANSI/ASME B1.7M-1984 - *NOMENCLATURE, DEFINITIONS, AND LETTER SYMBOLS FOR SCREW-THREADS* [1] For example: *form of thread* - the form of a thread is its profile in an axial plane for a length of one pitch of the complete thread.

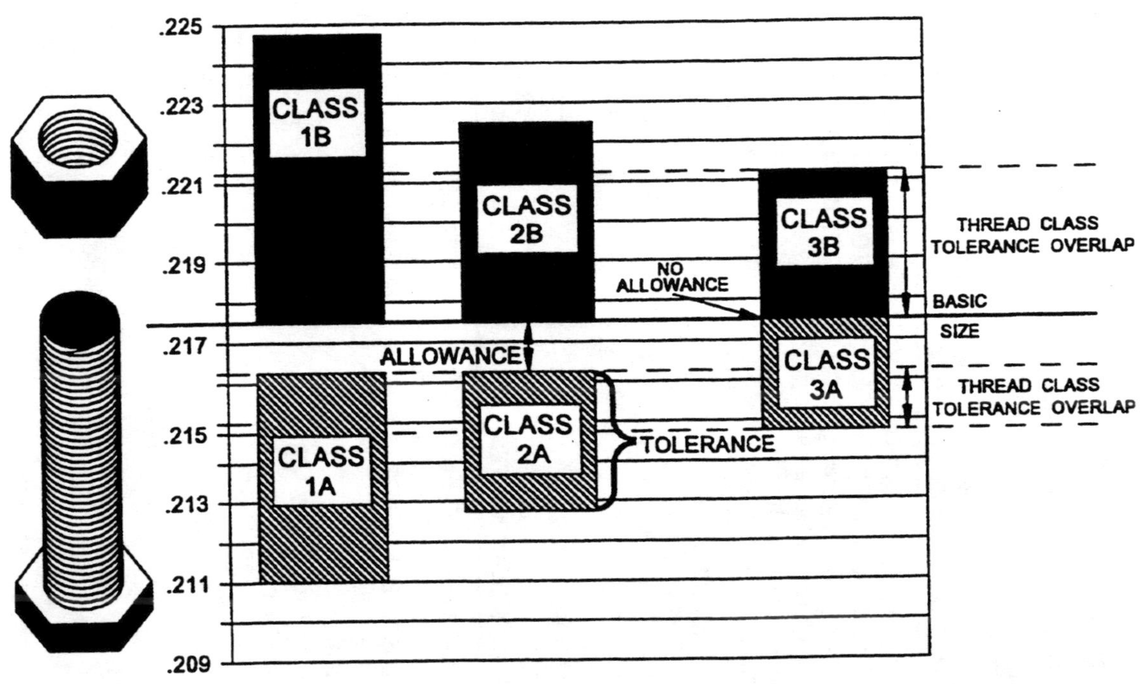

I-3. Distinctions among the three basic screw-thread classes based on tolerances and allowance. (Thread class 3A/3B has no allowance.) [4]

TABLE I-1
ABBREVIATED CHRONOLOGY OF DOMESTIC SOURCES OF SCREW-THREAD STANDARDIZATION DOCUMENTS

<table>
<tr><th colspan="2">AMERICAN NATIONAL STANDARDS</th><th colspan="2">FEDERAL PROCUREMENT STANDARDS</th><th colspan="2">MILITARY SPECIFICATIONS</th></tr>
<tr><th>SOURCE OF STANDARDIZATION DOCUMENTS</th><th>YEAR</th><th>SOURCE OF STANDARDIZATION DOCUMENTS</th><th>YEAR</th><th>SOURCE OF STANDARDIZATION DOCUMENT</th><th>YEAR</th></tr>
<tr><td>AMERICAN NATIONAL STANDARDS COMMITTEE B1</td><td>1918</td><td>NATIONAL SCREW-THREAD COMMISSION AUTHORIZED BY U.S. CONGRESS. An interagency standard developed which eventually took the form of a handbook.
CONSENSUS DOCUMENT</td><td>1918
TO
AROUND
1939</td><td>**CUSTODIAN:** The U.S. Air Force is the Defense Department's executive agent for Class 3 threaded fasteners.

NON-CONSENSUS DOCUMENTS</td><td></td></tr>
<tr><td rowspan="3">ASME STANDARDS COMMITTEE B1
CUSTODIAN:
ASME
345 East 47th Street
New York, New York 10017
(212) 705-7099
CONSENSUS DOCUMENTS</td><td rowspan="3">1920
TO
PRESENT</td><td rowspan="3">NATIONAL BUREAU OF STANDARDS — HANDBOOKS H25 and H28. NBS assumed custodianship of the interagency standard in 1939.

NBS H25/1939 (Ref. 2)
NBS H28/1942-1969 (Ref. 5)
CONSENSUS DOCUMENTS</td><td rowspan="3">AROUND
1939
TO
1977</td><td>MIL-S-7742A</td><td>2 DEC.
1959</td></tr>
<tr><td>MIL-S-8879 (ASG)</td><td>21 SEPT.
1960</td></tr>
<tr><td>MIL-S-8879A</td><td>8 DEC.
1965</td></tr>
<tr><td rowspan="6">SAE Thread Standards

CUSTODIAN:
SAE Committee E25
Society of Automotive Engineers
400 Commonwealth Drive
Warrendale, PA 15096-0001
(412) 776-4841

CONSENSUS DOCUMENTS</td><td rowspan="6"></td><td rowspan="6">FED-STD-H28 (23 Sections) (Ref. 6)
NBS HANDBOOK H28 was converted into FED-STD-H28 in 1977.

CUSTODIAN:
Defense Industrial Supply Center
700 Robbins Avenue
Philadelphia, PA 19111-5096
(702) 252-7429

NON-CONSENSUS DOCUMENT</td><td rowspan="6">1977
TO
PRESENT</td><td>MIL-S-7742B</td><td>2 FEB.
1968</td></tr>
<tr><td>MIL-S-008879B</td><td>29 JULY
1988</td></tr>
<tr><td>MIL-S-007742C</td><td>29 JULY
1988</td></tr>
<tr><td>MIL-S-8879C (Ref. 7)</td><td>25 JULY
1991</td></tr>
<tr><td>MIL-S-7742D (Ref. 8)
NOTE: Inactive for new design after 31 December 1991. For new design use MIL-S-8879</td><td>25 JULY
1991
TO
PRESENT</td></tr>
</table>

States. There are four custodians: 1) the Standards Committee B1 of the American Society of Mechanical Engineers, 2) the Society of Automotive Engineers Committee E25, 3) the Defense Industrial Supply Center, and 4) the United States Air Force for military thread specifications.

5. Inspection Standards for Threaded Products

Screw-thread inspection standards specify methodologies for examining threaded products. These methodologies are termed "inspection methods" or "gaging systems".

An inspection standard and an inspection method/gaging system must be designated in an engineering drawing or other contractual document or contract clause. The designation is made by the designer/procurer of a threaded product and reflects the application needs.

6. Degree of Inspection Severity

Inspection standards for threaded products offer a wide range of inspection severity. The degree of inspection severity applied to threaded products is chosen by the designer/procurer of the product by selecting an inspection method/gaging system. The wide range of inspection severity in the thread inspection standard is further augmented by a wide range of gages and measuring equipment from which the designer/procurer can select. (See Sections II-C and III-C.)

7. Dimensional Conformance and Acceptability

A screw-thread inspection standard may address *dimensional conformance* or it may address *acceptability*.

> *Dimensional conformance* is an indication that thread elements and thread characteristics of a threaded product have sizes within the maximum and minimum limits of size which are specified in the applicable thread standard.
>
> *Acceptability* is an indication that the threaded product has passed the inspection requirements specified in the selected gaging system in the designated inspection standard.

ASME B1.3M-1992 [10] is titled, *Screw Thread Gaging Systems for Dimensional Acceptability — Inch and Metric Screw Threads (UN, UNR, UNJ, M, and MJ)*. The focus of this document is on *acceptability*.

8. Historical Note — Origin of ASME B1.3M

ASME B1.3M originated in a controversy which occurred about twenty-five years ago. The controversy was about gages for inspecting pitch diameter. ASME B1.1-1960 contained a paragraph "22" which specified that the threads of a product would be considered *acceptable* if they passed functional diameter inspections made by **GO** and **NOT GO** gaging. However, thread class 3A had the additional requirement that the minimum pitch diameter must also be inspected because it was widely recognized that **NOT GO** gaging could not fully inspect for minimum pitch diameter.

In 1969, ANSI B1.1-1960 came under sharp criticism because paragraph "22" did not include indicating gages which had been developed by 1969 and which could be used to inspect the pitch diameter of a threaded product. Nevertheless, ASME B1.1 was issued at that time without the acceptability paragraph alluding to indicating gages. In response to the criticism, a subcommittee was formed which was charged with creating a document which could be used by procurers/designers to specify acceptability criteria to satisfy a broad range of product applications.

ASME B1.3M — *Screw-Thread Gaging Systems for Dimensional Acceptability — Inch and Metric Screw-Threads (UN, UNR, UNJ, M, and MJ)* resulted from the work of this subcommittee. The first edition of this standard, ANSI B1.3, was issued in June, 1979. Most of this edition and of subsequent editions was developed by a consensus committee led by gage manufacturers and product end-users. The intention was to include known thread inspection gages and measuring equipment and to require the procurer/designer to specify the degree of inspection severity. This intention was met.

Today's document, ASME B1.3M-1992 [10], offers three inspection methods, namely, Systems 21, 22, and 23. System 21 inspects the fewest thread elements and thread characteristics, whereas System 23 inspects the most thread elements and thread characteristics. ASME B1.3M-1992 lists known gages and measuring equipment for inspection of thread elements and thread characteristics of a threaded product.

The procurer/designer selects an inspection system in ASME B1.3M based on the control of thread elements/characteristics needed to meet the demands of the application. The gages and measuring equipment used for thread inspection are selected by the inspection department at the product vendor and by the quality control program at the procurer or designer end of the transaction. Such quality programs have been proliferating and intensifying in the last twenty years.

ASME B1.3M offers considerable flexibility for thread inspection. For example, sub-paragraph 7(b) of ASME B1.3M-1992 [10] has provision to modify any of the three inspection systems so that: 1) thread elements or thread characteristics in addition to those already specified by System 21, System 22, or System 23, can also be specified; or 2) thread elements or thread characteristics specified by System 21, System 22, or System 23 can be deleted from the specified inspection.

An integrated package for thread manufacture and inspection involves: 1) a screw-thread standard, e.g., ASME B1.1-1989 [3], 2) a standard for gages and gaging, e.g., ANSI/ASME B1.2-1983 [12], and 3) a screw-thread acceptability standard, e.g., ASME B1.3M-1992 [10]. These standards can best be understood if a nomenclature standard is also in hand, e.g., ANSI/ASME B1.7M-1984 [1]. The standard for gages and gaging, and the standard for gaging systems, are automatically invoked when the screw-thread standard itself is designated. Together, these standards set the stage for manufacture of a screw-thread product which meets the required product performance.

Years ago when metrication was being promoted, a metric thread series was defined by ASME in ASME B1.18M-1982 — *Metric Screw-Threads for Commercial Mechanical Fasteners - Boundary Profile Defined* (withdrawn) [12], and in ASME B1.19M-84 — *Gages for Metric Screw-Threads for Commercial Mechanical Fasteners - Boundary Profile Defined* (withdrawn) [13]. These metric threads were primarily selected diameter-pitch combinations but were defined without a pitch diameter. The thread form was defined in ASME B1.18 [12] by major diameter and crest width, minor diameter, and root contour. Acceptance of these threads was by **GO** and **NOT GO** gaging defined in B1.19M [13]. The **NOT GO** gages contacted the thread flanks at two locations. However, this approach to thread inspection did not receive wide acceptance, and these two thread documents were withdrawn.

II. THREADED PRODUCT STANDARDS AND GAGING SYSTEMS

A. Overview of Thread and Gaging Standards

1. Thread Standards

General interchangeability of threads is assured through the standardization of thread form, diameter-pitch combinations, and limits of size, as set forth in applicable thread standards.

This study focuses on *unified inch screw-threads*, as defined in the thread standard, ANSI/ASME B1.1-1989, *Unified Inch Screw-Threads (UN and UNR Thread Form)* [3]. The basic profile for UN and UNR 60° screw-threads is shown in Figure I-2 of this report. Diametral dimensions of this thread form appear in Figures II-1A and II-1B of this report.

2. Definition and Purpose of Gaging

Gaging is the process of inspecting a threaded product using gages specified in a standardization document. An example of such a standard is ANSI/ASME B1.2-1983, *Gages and Gaging for Unified Inch Screw-Threads* [11].

The purpose of gaging[3] is to assure that standardized screw thread products from different sources, which have passed standardized inspections, are controlled to the extent specified by the procurer/designer. Control sought by the procurer/designer may include any or all of the following: 1) acceptable fit (assembly); and 2) size (conformance to dimensional specifications). The inspection process may be qualitative (*limit* or *attributes* inspection) or quantitative (*size* inspection). The procurer/designer should be aware that the standards/specifications for inspection of screw thread products do not address the mechanical performance of assemblies made using threaded products which have passed inspection.

Gaging is always carried out to determine the *acceptability* of a product screw thread. Such inspection involves qualitative determination of: 1) the form of thread, and 2) boundaries or limits of dimensions of specified product thread elements or characteristics.

Gaging is applied at various times in the life of a threaded product. For example: 1) gaging is used in the set-up stage of high-volume manufacturing processes for screw-thread products to establish process control. Subsequently, gaging is periodically used to inspect the stream of threaded products being manufactured. For one-of-a-kind threaded products, gaging is used in the set-up and creation stages of the product to assure acceptability; 2) Gaging may be used by a procurer/designer to inspect incoming products and product inventory; and 3) Gaging may be used to certify that threaded products conform to requirements in a procurement specification or on an engineering drawing.

---

[3]The following statement about the purpose of gaging appears in NBS Handbook H25-1939 [2]:

> "The final results sought by gaging are to secure interchangeability, that is, the assembly of mating parts without selection or fitting of one part to another, and to insure that the product conforms to the specified dimensions within the limits of variation establishing the closest and loosest conditions of fit permissible in any given case, as provided for in the foregoing specifications."

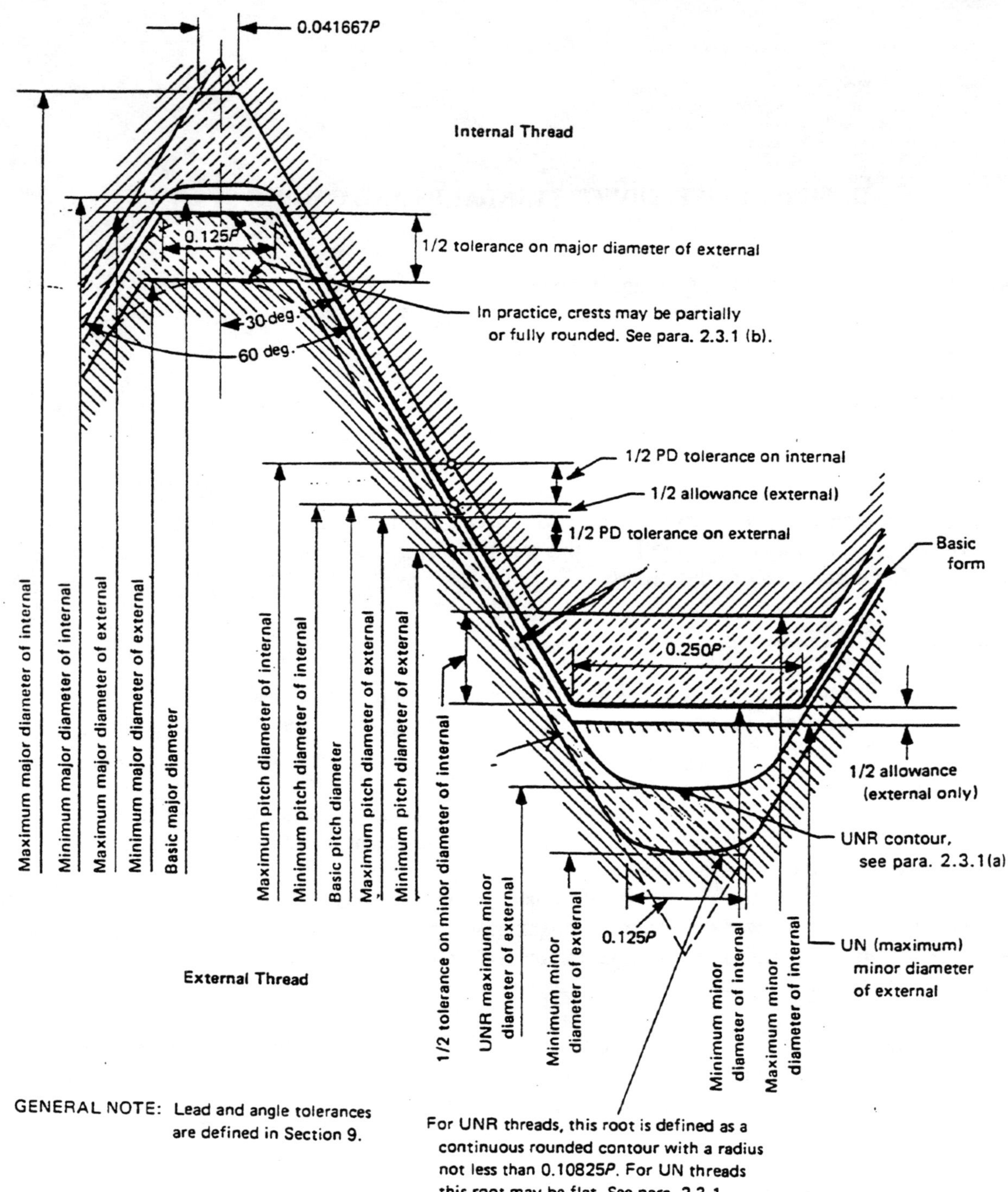

II-1A. SCREW THREAD CLASSES 1A, 2A, 1B and 2B [3].

Highly-magnified longitudinal cross-section view of internal and external threads, showing Basic Pitch Diameter at mid-point of one thread flank. Also shown are radial dimensions of elements of the internal and external threads. Tolerances and allowance shown are derived from radial dimensions.

The *thread-definition envelope (tolerance zone)* for the internal and external thread appears as a shaded region along each thread's crest, flank, and root.

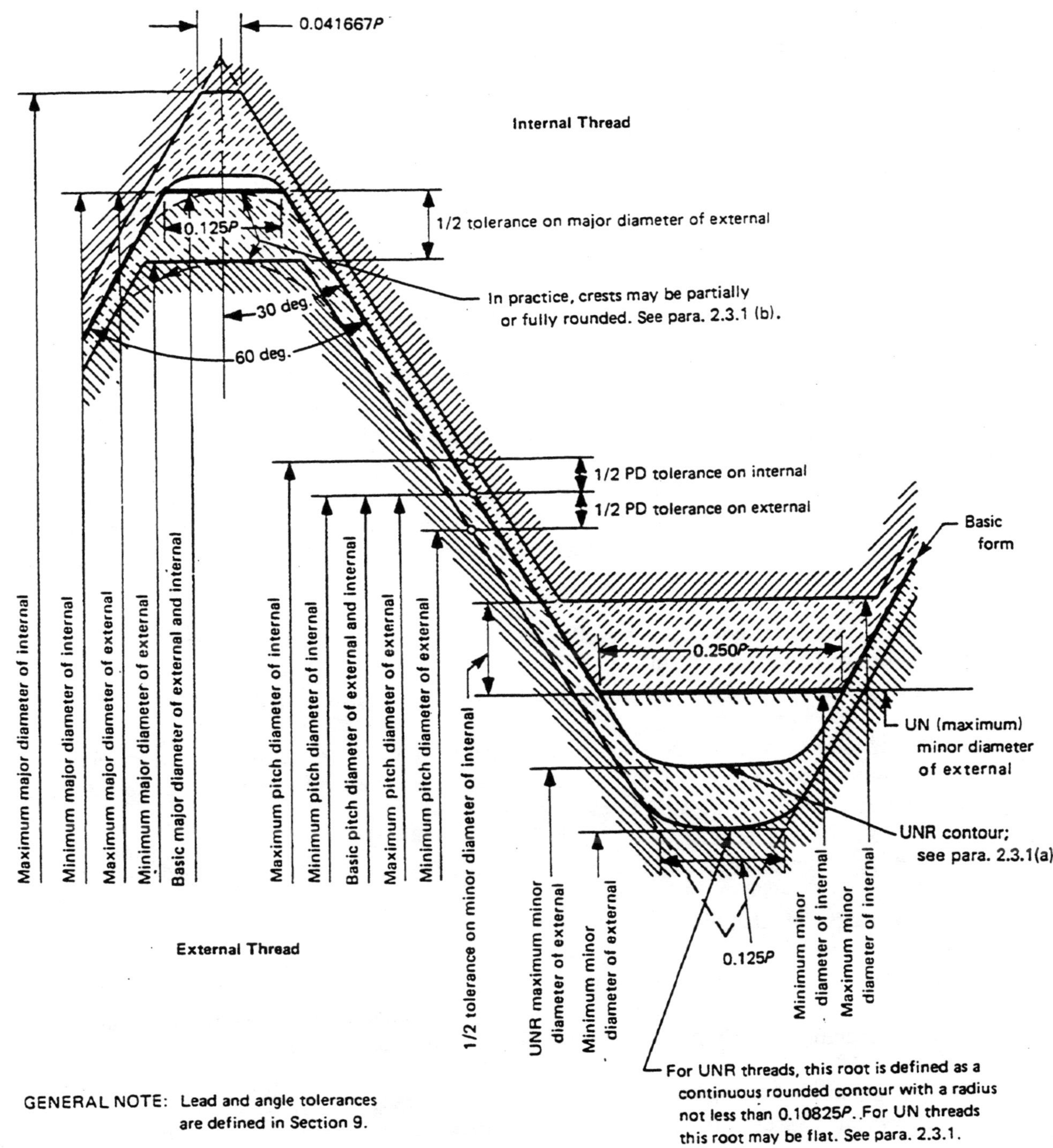

II-1B. SCREW THREAD CLASSES 3A and 3B [3].

Highly-magnified longitudinal cross-section view of internal and external threads, showing Basic Pitch Diameter at mid-point of one thread flank. Also shown are radial dimensions of elements of the internal and external threads. The tolerances shown are derived from radial dimensions. Thread Class 3 has no allowance.

The *thread-definition envelope (tolerance zone)* for the internal and external thread appears as a shaded region along each thread's crest, flank, and root.

3. Thread-Definition Envelope

The envelope defining a thread (which is a helix) includes the acceptable range of *thread assembly diameters*[4]. The inner and outer radial boundaries of the helix are given by the minimum and maximum pitch diameter, respectively, of a screw-thread. The longitudinal axis of this helix coincides with the longitudinal axis of the screw thread, and the pitch of this helix is identical to the pitch of the screw-thread.

The thread-definition envelope can be seen in longitudinal cross section in Figure II-1A (Classes 1A, 1B, 2A, 2B) and Figure II-1B (Classes 3A and 3B). These diagrams are from pages 4 and 5, respectively, of the thread standard, ASME B1.1-1989, *Unified Inch Screw-Threads (UN and UNR Thread Form)* [3]. The envelope appears as a cross-hatched region along the thread flank in each figure.

Envelopes for Class 1A/1B and Class 2A/2B internal and external threads are centered around the label, "Basic Pitch Diameter," in Figure II-1A. Flank contact does not occur for concentric threads of Classes 1A/1B and 2A/2B due to the *allowance*. Allowance is specified only for Class 1A and 2A external threads. Allowance is defined qualitatively in ANSI/ASME B1.7-1984 (page 1) [1]. In essence, allowance is the difference between the internal-thread minimum-pitch-diameter and external-thread maximum-pitch-diameter.

Envelopes for Class 3A external and Class 3B internal threads are centered around the label, "Basic pitch diameter of external and internal," in Figure II-1B. Flank contact for the combination of internal-thread minimum-pitch-diameter and external-thread maximum-pitch-diameter is evident in Figure II-1B.

Pitch diameter tolerances for internal and external threads define the dimensions of the envelope in the radial direction for both internal and external threads. The Basic Pitch Diameter for Classes 3A and 3B is equal to both of the following diameters:

-the minimum pitch diameter of the internal thread, and

-the maximum pitch diameter of the external thread.

The maximum pitch diameter of the internal thread extends radially outward from the Basic Pitch Diameter and is located at a radial distance from the centerline equal to one-half the maximum pitch diameter specified for the internal thread in the applicable thread standard.

The minimum pitch diameter of the external thread extends radially inward from the Basic Pitch Diameter and is located at a radial distance from the centerline equal to one-half the minimum pitch diameter specified for the external thread in the applicable thread standard.

The Basic Pitch Diameter for Classes 1A, 2A, 1B, and 2B is equal to the minimum pitch diameter of the internal thread.

---

[4]See ANSI/ASME B1.7M, *NOMENCLATURE, DEFINITIONS, AND LETTER SYMBOLS FOR SCREW-THREADS* [1] (pg 6/col.1 for definition of *thread assembly diameter.*)

The maximum pitch diameter of the internal thread extends radially outward beyond the Basic Pitch Diameter and is located at a radial distance from the centerline equal to one-half the maximum pitch diameter specified for the internal thread in the applicable thread standard.

The maximum pitch diameter of the external thread is located radially inward from the Basic Pitch Diameter a distance equal to one-half of the allowance.

The minimum pitch diameter of the external thread extends radially inward from the maximum pitch diameter of the external thread and is located at a radial distance from the centerline equal to one-half the minimum pitch diameter specified for the external thread in the applicable thread standard.

The thread-definition envelope is sometimes termed the *tolerance zone*, which is defined in ANSI/ASME B1.7M [1] as ... "the zone between the maximum and minimum limits of size." Inspection of the thread definition envelope is difficult, as discussed in Appendix A.

4. Importance of Thread-Definition Envelope

The thread-definition envelope is a very important concept in screw-thread design and construction. The thread-definition must include a tolerance on pitch diameter due to radial uncertainty of pitch diameter arising from variations in thread elements/thread characteristics which occur during manufacture. Such variation is the source of radial uncertainty of pitch diameter, which is taken into account by placing tolerances on the pitch diameter.

The ranges for maximum and minimum pitch diameters, i.e., the tolerances on pitch diameters, vary with thread size and class. ASME B1.1-1989 (Table 3A, pages 23-46) [3] shows the tolerances for standard series threads (UN/UNR).

B. Screw-Thread Gaging Systems

Inspection of threads by a screw-thread gaging system is the critical mechanism by which standardized threaded products are evaluated in order to assure success in an assembly. When a designer/procurer specifies a thread class in the applicable screw-thread standardization document, a screw thread inspection standard may be added to the screw thread designation. An example of this point appears in ASME B1.1-1989 (Section 6, page 12) [3]..

A screw-thread gaging system is a very powerful concept, for it specifies: 1) the gage standard — which in turn specifies: a) the design of gages and measuring equipment (including tolerances) to be used for inspection, and b) the *calibration* requirements for the gages and measuring equipment; 2) the thread characteristics and/or thread elements to be inspected; and 3) the conditions for *acceptability* or *dimensional conformance*.

Screw-thread gaging systems (also called *inspection methods*) are specified in screw-thread inspection standards/specifications. An inspection document must be designated in a procurement contract, or the designer/procurer may select one.

The first formalized inspection methods to appear in a screw thread standardization document were specified in 1973 in two Military Specifications, namely, MIL-S-7742 and MIL-S-8879.

C. Standardization Documents Which Specify Screw-Thread Gaging Systems

Standardization documents which specify screw-thread gaging systems or inspection methods are found in American National Standards, Federal Procurement Standards, Military Specifications, and ISO standards. Table II-1 lists current standardization documents which specify screw-thread gaging systems. The only current domestic standardization documents which designate more than one screw-thread gaging system are ASME B1.3M-1992 [10] and FED-STD-H28/20B-1994 [14][5].

Tables II-2A and II-2B summarize the evolution of gaging systems and inspection methods in domestic standardization documents. The evolution seems to have followed a give-and-take between various interested parties — specifically, custodians and caretakers for American National Standards, for Federal Procurement Standards, and for Military Specifications.

A choice of three inspection methods/gaging systems first appeared in domestic standardization documents in the March 15, 1973 amendments to MIL-S-8879A-1965 and MIL-S-7742B-1968.

D. Two Inspection Standards Specifying Systems 21, 22, and 23

Two different thread-inspection standards listed in Tables II-2A and II-2B of this report, ASME B1.3M-1992 [10] and FED-STD-H28/20B-1994 [14], refer to screw-thread gaging systems. The current version of each of these thread-inspection standards is titled:

ASME B1.3M-1992:	*Screw-Thread Gaging Systems for Dimensional Acceptability — Inch and Metric Screw-Threads (UN, UNR, UNJ, M, AND MJ)*
FED-STD-H28/20B-1994:	*Inspection Methods for Acceptability of UN, UNR, UNJ, M, and MJ Screw-Threads*

The three gaging systems in each standard[6] are termed, *System 21, System 22,* and *System 23.* One important difference between these two inspection standards is the focus of the inspection methodology. ASME B1.3M [10] focuses on *the threaded product being inspected,* whereas FED-STD-H28/20 [14] focuses on *the gages and measuring equipment used for inspection.* ASME B1.3M [10], therefore, appears to be more inclusive and more flexible than FED-STD-H28/20 [14].

ASME B1.3M-1992 emphasizes the *thread elements and thread characteristics of the threaded product being inspected,* and lists the gages and measuring equipment suitable for inspecting each element/characteristic.

---

[5]Two provisional National Aerospace Standards, NAS 8879 and NAS 7742, specify screw-thread gaging systems and inspection methods. These standardization documents (distributed, but not adopted) were intended to serve as civilian alternatives to the military standards on which they were based, namely, MIL-S-8879A and MIL-S-7742B and their associated amendments. The standards are not American National Standards. Rather, they were issued in 1991 by the National Aerospace Standards Committee of the Aerospace Industries Association (1250 Eye St., NW, Suite 1100, Washington, DC/202-371-8433). NAS8879 and NAS7742 were never active screw-thread standards. Both remain in provisional status.

[6]ASME B1.3M-1992 [10] offers a fourth screw-thread gaging system for metric threads, System 21A. This gaging system is soon to be discontinued. System 21A was incorporated in ASME B1.3M to accommodate inspection of metric threads with metric thread gages and measuring equipment, as specified, respectively, in ASME B1.18M-82 - *Metric Screw-Threads for Commercial Mechanical Fasteners - Boundary Profile Defined* (withdrawn) [12], and in ASME B1.19M-84 - *Gages for Metric Screw-Threads for Commercial Mechanical Fasteners - Boundary Profile Defined* (withdrawn)[13].

TABLE II-1
CURRENT STANDARDIZATION DOCUMENTS
WHICH SPECIFY SCREW-THREAD GAGING SYSTEMS

CUSTODIAN	DOCUMENT	COMMENTS
ASME (New York) **CONSENSUS**	**ASME B1.3M-1992** Screw-Thread Gaging Systems For Dimensional Acceptability - Inch and Metric Screw-Threads (UN, UNR, UNJ, M, and MJ) [10]	Specifies three screw-thread gaging systems. SYSTEM 21; SYSTEM 22; SYSTEM 23
DEFENSE INDUSTRIAL SUPPLY CENTER (Philadelphia) **NON-CONSENSUS**	**FED-STD-H28/20B FEDERAL STANDARD** - SCREW-THREAD STANDARDS FOR FEDERAL SERVICES - SECTION 20 - INSPECTION METHODS FOR ACCEPTABILITY OF UN, UNJ, M AND MJ SCREW THREADS (10 March 1994) [14]	Specifies three screw-thread gaging systems. References ASME B1.3M-1992 [10] SYSTEM 21; SYSTEM 22; SYSTEM 23
U.S. AIR FORCE (Washington, DC) **NON-CONSENSUS**	**MIL-S-8879C** SCREW-THREADS, CONTROLLED RADIUS ROOT WITH INCREASED MINOR DIAMETER, GENERAL SPECIFICATION FOR (25 July 1991) [7]	Specifies two thread application categories - "Safety Critical Threads" and "Other Threads." Safety Critical - specifies **eleven** thread characteristics for inspection and 100% product inspection. Other - specifies **five** thread characteristics for inspection. References the following thread inspection standards: **FED-STD-H28/20,** and **ASME B1.3M**
U.S. AIR FORCE (Washington, DC) **NON-CONSENSUS**	**MIL-S-7742D** · SCREW-THREADS, STANDARD, OPTIMUM SELECTED SERIES: GENERAL SPECIFICATION FOR (25 July 1991) [8]	Same comments as for MIL-S-8879C. Inactive for new design after 31 December 1991.
ISO/TC1 Geneva, Switzerland **NON-CONSENSUS**	**ISO 1502** ISO GENERAL PURPOSE METRIC SCREW-THREADS GAUGING (15 July 1978) [9]	Specifies one screw-thread gaging system. (Similar to System 21)

TABLE II-2A
CHRONOLOGY OF SCREW-THREAD INSPECTION SYSTEMS

SOURCE DOCUMENT FOR SCREW-THREAD INSPECTION SYSTEM	IDENTITY OF SCREW-THREAD INSPECTION SYSTEM AND DATE SOURCE DOCUMENT WAS FIRST APPROVED			
	NBS HANDBOOK H28 (5)	MILITARY SPECS METHODS A-B-C	SYSTEMS 21, 22, 23	
			ASME B1.3M (10)	FED-STD-H28/20 (14)
NBS HANDBOOK H28 (NOTE 1)	1942-1969			
MILITARY SPECS MIL-S-8879 & 7742 (NOTE 2) (7,8) **(NON-CONSENSUS)**		15 MARCH 1973		
ASME B1.3/B1.3M (NOTE 3) **(CONSENSUS)**			21 JUNE 1979	
FED-STD-H28/20 (NOTE 4) **(NON-CONSENSUS)**				30 SEPTEMBER 1981

(1) Screw thread gaging systems were not explicitly specified, although gages which are included in Systems 21, 22 and 23 were described in the 1969 edition. Any inspection system could be used unless otherwise specified. The 1957 edition described gages in the current Systems 21 and 22 only, while previous editions only described gages in the current System 21.

(2) The designer/procurer was offered a choice of three inspection methods: A, B and C. These methods were similar to the current Systems 21, 22 and 23, respectively. Unless otherwise specified, Method A was to be used for internal threads and B for external.

The focus of inspection was on specific *thread elements* and *thread characteristics*. *Length of thread engagement during inspection* was explicitly introduced into the inspection methodology. The phrase, "screw thread gaging system", was not used.

The 25 July 1991 editions (See Table 1) of these two military specifications:

- deleted inspection Methods A, B, and C. Two inspection standards, FED-STD-H28/20 and ASME B1.3M were incorporated by reference to provide inspection methodologies.

- introduced two "Application Categories" — *Safety Critical Thread* and *Other Thread*, with thread characteristics to be checked, equivalent to ASME B1.3M — Systems 23 and 22, respectively, unless otherwise specified.

TABLE II-2B
CHRONOLOGY OF SCREW-THREAD INSPECTION SYSTEMS

SOURCE DOCUMENT FOR SCREW-THREAD INSPECTION SYSTEM	IDENTITY OF SCREW THREAD INSPECTION SYSTEM AND DATE SOURCE DOCUMENT WAS FIRST APPROVED			
	NBS HANDBOOK H28 (5)	MILITARY SPECS METHODS A-B-C	SYSTEMS 21, 22, 23	
			ASME B1.3M (10)	FED-STD-H28/20 (14)
NBS HANDBOOK H28 (NOTE 1)	1942-1969			
MILITARY SPECS MIL-S-8879 & 7742 (NOTE 2) (7,8) **(NON-CONSENSUS)**		15 MARCH 1973		
ASME B1.3/B1.3M (NOTE 3) **(CONSENSUS)**			21 JUNE 1979	
FED-STD-H28/20 (NOTE 4) **(NON-CONSENSUS)**				30 SEPTEMBER 1981

(3) The designer/procurer was offered a choice of three inspection methods, termed System 21, System 22, and System 23. Severity of inspection increased from System 21 to System 23. The focus of inspection was on specific *thread elements* and *thread characteristics*. The designer/procurer was offered a choice of a wide range of gages and measuring equipment and was required to specify the selected gaging system.

Length of thread engagement during inspection is not explicitly mentioned for inspections specified in ASME B1.3M-1992, but is included in the gage design information in the ASME standards on gages and gaging (B1.2, B1.16M, B1.22M).

Visual inspection of product threads is an explicit requirement of all inspections specified in ASME B1.3M-1992.

(4) The designer/procurer was offered a choice of three inspection methods, termed System 21, System 22, and System 23. Severity of inspection increased from System 21 to System 23. Inspection focused on *gages and measuring equipment, calibration of gages, and gaging methodology.*

Length of thread engagement during inspection is not explicitly mentioned for inspections specified in FED-STD-H28/20B dated March 10, 1994, but is required in accordance with ASME B1.3M which is incorporated by reference. (See H28/20B para. 5.2)

Visual inspection of product threads is not explicitly mentioned for inspections specified in FED-STD-H28/20B dated March 10, 1994.

FED-STD-H28/20B-1994 emphasizes the *gages and measuring equipment* used for inspection and also emphasizes *the calibration requirements* for the thread elements and thread characteristics of the thread gages themselves. The calibration and operation of measuring equipment is specified to be in accord with instructions from the manufacturers of measuring equipment.

Another important difference between ASME B1.3M-1992 [10] and FED-STD-H28/20B-1994 [14] is the amount of guidance which each document provides. FED-STD-H28/20B-1994 offers the designer/procurer more explanatory material for government applications while ASME B1.3M-1992 is directed toward more general applications. For example, in the matter of selecting a gaging system which is suitable for the intended application of a threaded product, FED-STD-H28/20B-1994 (pages 4 and 5) suggests conditions under which each gaging system might be suitable, whereas ASME B1.3M-1992 does not offer such guidance.

Two examples of the guidance provided in FED-STD-H28/20B-1994 for selection of a gaging system follow:

¶5.1.3.2 System 21 is suggested for use under <u>any one</u> (or more) of the following conditions:

(a) "Where the threads of the product do not need specific mechanical strength properties, or where mechanical strength requirements are not specified for the product threads either by material strength and dimensional limits or by testing strength of the threads."

(c) "For standard, off-the-shelf, general application fasteners when considered acceptable."

¶5.1.5.2 System 23 is suggested for use under <u>any one</u> (or more) of the following conditions:

(a) "When thread element control is required to determine the extent of deviation in any of the elements of the thread; normally, special applications."

E. Military Specifications Using Systems 21, 22, and 23

The two inspection standards, ASME B1.3M [10] and FED-STD-H28/20B [14], are incorporated by reference into the most recent editions (1991) of the two Military Specifications listed in Table II-1, namely, MIL-S-8879C [7] and MIL-S-7742D [8]. Hence, military procurers/designers are empowered to utilize inspections by Systems 21-23. The level of inspection severity is determined by the "Application Category" described in Table II-1 of this report.

III. COMPARISONS OF SYSTEMS 21, 22, AND 23

A. Comparing Systems 21, 22, and 23 in ASME B1.3M-1992 with Each Other:

Systems 21, 22, and 23 are unique. They are the primary screw-thread gaging systems which can be used domestically for inspecting threaded products. Considering the uniqueness of Systems 21-23, it is instructive to compare what is accomplished by inspections carried out under each system.

Table III-1A and Table III-1B in this report, derived from Tables 1 and 3 in ASME B1.3M-1992 [10], compare the inspections of thread elements and thread characteristics which are specified by Systems 21-23 in ASME B1.3M-1992. These tables show at a glance:

- how much choice and flexibility a designer/procurer has when using ASME B1.3M-1992 [10] to select a gaging system for threaded product inspections.

- that *size* or *limit* inspections can be selected by the designer/procurer of threaded products.

- that System 21 inspects two thread characteristics and one element termed:

  1) "Maximum Material - **GO**"

  2) "**NOT GO** Functional Diameter"

  3) "Major/Minor Diameter" (external/internal)

- that System 21 is the only system which inspects the thread characteristic termed "**NOT GO** Functional Diameter".

In addition to the information summarized in Tables III-1A and III-1B, text in each inspection standard describes important aspects of the inspection system. For example:

ASME B1.3M-1992, ¶4(b)(1): *System 21.* Provides for interchangeable assembly with functional size control at the maximum material limits within the length of standard gaging elements, and also provides for control of the characteristics identified as **NOT GO** functional diameters.

FED-STD-H28/20B-1994 ¶5.1.3.1: System 21 provides for interchangeable assembly with functional size control at the maximum material limits within the length of standard gaging elements; and also provides for control of characteristics identified as **NOT GO** functional diameters or as **HI** (Internal) and **LO** (External) functional diameters. These functional gages provide some control at the minimum material limit when there is little variation in thread form characteristics such as lead, flank angle, taper and roundness

TABLE III-1A EXTERNAL PRODUCT THREAD
COMPARISON OF THREAD CHARACTERISTICS AND THREAD ELEMENTS
INSPECTED BY THREE SCREW-THREAD GAGING SYSTEMS IN ASME B1.3M-1992 [10]

SCREW-THREAD GAGING SYSTEM	MINIMUM MATERIAL			MAXIMUM MATERIAL GO	NOT GO FUNCTIONAL DIAMETER	MAJOR DIAMETER	COMMENTS
	PITCH DIAMETER	THREAD GROOVE DIAM.	NOT GO FN.D& LEAD&FLANK				
21				*INSPECTED* WITH GO-GAGE FUNC. LIMIT 1.1; 2.1; 2.3; 4.1; 4.3. FUNC. SIZE 4.1; 4.3.	*INSPECTED W.* NOT-GO GAGE FUNC. LIMIT 1.2; 2.2; 2.4; 4.1; 4.3; 6. FUNC. SIZE 4.1; 4.3; 6.	*INSPECTED* LIMIT 3.1(a); 3.1(b); 3.2; 3.4; 5.1; 9; 14. SIZE 5.1; 9; 14; 17; 18.	*SYSTEM 21 REQUIRES NO OTHER INSPECTIONS.* NOTE A: THE PRIMARY RESULT OF SYSTEM 21 INSPECTION IS DETERMINATION OF WHETHER OR NOT THREADED PRODUCT WILL ASSEMBLE. NOTE B: THE ONLY THREAD ELEMENT INSPECTION REQUIRED BY SYSTEM 21 IS MAJOR DIAM. NOTE C: *MINIMUM MATERIAL* CANNOT BE FULLY INSPECTED BY SYSTEM 21.
22	*INSPECTED* LIMIT 2.5; 4.5; 7. SIZE 4.5; 7; 17; 18.	*INSPECTED* LIMIT 2.6; 4.6; 8. SIZE 4.6; 8; 18.	*INSPECTED** FUNC. LIMIT 1.2; 2.2; 2.4; 4.1; 4.3; 6. SIZE - LEAD 4.8; 9; 11; 12; 13; 17; 18. SIZE - FLANK 4.8; 9; 10; 18.	*INSPECTED* WITH GO-GAGE FUNC. LIMIT 1.1; 2.1; 2.3; 4.1; 4.3 FUNC. SIZE 4.1; 4.3.		*INSPECTED* LIMIT 3.1(a); 3.2; 3.4; 5.1; 9; 14. SIZE 5.1; 9; 14; 17; 18.	SYSTEM 22 REQUIRES FURTHER INSPECTIONS. (SEE TABLE III-1B)
23	*INSPECTED* LIMIT 2.5; 4.5; 7. SIZE 4.5; 7; 17; 18.	*INSPECTED* LIMIT 2.6; 4.6; 8. SIZE 4.6; 8; 18.		*INSPECTED* WITH GO GAGE FUNC. LIMIT 1.1; 2.1; 2.3; 4.1; 4.3. FUNC. SIZE 4.1; 4.3.		*INSPECTED* LIMIT 3.1 (a); 3.2; 3.4; 5.1; 9; 14. SIZE 5.1; 9; 14; 17; 18.	SYSTEM 23 REQUIRES FURTHER INSPECTIONS. (SEE TABLE III-1B)

NOTE 1: The numbers in each column represent *thread gages* and *measuring equipment* specified in ASME B1.3M, Table 1. [10]
NOTE 2: ***LIMIT*** is inspected with a "thread limit gage" or with a "thread indicating gage"—***SIZE*** with a "thread indicating gage" or with other measuring equipment.
NOTE 3 *This combination of limit and size inspection is only by agreement between purchaser and supplier. (ASME B1.1-1, 1989, ¶9.1.2, page 53) [3]

TABLE III-1B EXTERNAL PRODUCT THREAD
COMPARISON OF THREAD CHARACTERISTICS AND THREAD ELEMENTS
INSPECTED BY THREE SCREW-THREAD GAGING SYSTEMS IN **ASME B1.3M-1992** [10]

SCREW-THREAD GAGING SYSTEM	MINIMUM MATERIAL			MAXIMUM MATERIAL GO	NOT GO FUNCTIONAL DIAMETER	MAJOR DIAMETER	MINOR DIAMMETER (ROUNDED ROOT)	ROOT PROFILE	ROUNDNESS OF PITCH CYLINDER		TAPER OF PITCH CYLINDER	CUMMU. FORM VAR'N	LEAD INCL. HELIX VAR'N	FLANK ANGLE VAR'N	RUN-OUT MjDi PD VAR'N
	PITCH DIAMETER	THEAD GROOVE DIAM.	NOT GO FN.D& LEAD&FLANK						OVAL 180 DEG.	MULTILOBE 120 DEG.					
21				*INSPECTED* WITH GO-GAGE FUNC. LIMIT 1.1; 2.1; 2.3; 4.1; 4.3. FUNC. SIZE 4.1; 4.3.	*INSPECTED W.* NOT-GO GAGE FUNC. LIMIT 1.2; 2.2; 2.4; 4.1; 4.3; 6. FUNC. SIZE 4.1; 4.3; 6.	*INSPECTED* LIMIT 3.1(a); 3.1(b); 3.2; 3.4; 5.1; 9; 14. SIZE 5.1; 9; 14; 17; 18.									
22	*INSPECTED* LIMIT 2.5; 4.5; 7. SIZE 4.5; 7; 17; 18.	*INSPECTED* LIMIT 2.6; 4.6; 8. SIZE 4.6; 8; 18.	*INSPECTED** FUNC. LIMIT 1.2; 2.2; 2.4; 4.1; 4.3; 6. SIZE - LEAD 4.8; 9; 11; 12; 13; 17; 18. SIZE - FLANK 4.8; 9; 10; 18.	*INSPECTED* WITH GO-GAGE FUNC. LIMIT 1.1; 2.1; 2.3; 4.1; 4.3 FUNC. SIZE 4.1; 4.3.		*INSPECTED* LIMIT 3.1(a); 3.2; 3.4; 5.1; 9; 14. SIZE 5.1; 9; 14; 17; 18.	*INSPECTED* LIMIT 3.3; 3.5; 5.2; 9. SIZE 5.2; 9; 17; 18.	*INSPECTED* LIMIT SIZE 9; 10; 18.							
23	*INSPECTED* LIMIT 2.5; 4.5; 7. SIZE 4.5; 7; 17; 18.	*INSPECTED* LIMIT 2.6; 4.6; 8. SIZE 4.6; 8; 18.		*INSPECTED* WITH GO GAGE FUNC. LIMIT 1.1; 2.1; 2.3; 4.1; 4.3. FUNC. SIZE 4.1; 4.3.		*INSPECTED* LIMIT 3.1 (a); 3.2; 3.4; 5.1; 9; 14. SIZE 5.1; 9; 14; 17; 18.	*INSPECTED* LIMIT 3.3; 3.5; 5.2; 9. SIZE 5.2; 9; 17; 18.	*INSPECTED* LIMIT SIZE 9; 10; 18.	*INSPECTED* LIMIT 2.1; 2.2; 2.3; 2.4; 2.5; 2.6; 4.1; 4.3; 4.5; 4.6; 4.8; 6; 7; 8; 9; 16. SIZE 4.1; 4.3; 4.5; 4.6; 4.8; 6; 7; 8;9;16;17; 18.	*INSPECTED* LIMIT 4.1; 4.3; 4.5; 4.6; 4.8; 6; 7; 8; 9; 16; 17; 18. SIZE 4.1; 4.3; 4.5; 4.6; 4.8; 9; 16; 18.	*INSPECTED* LIMIT 2.2; 2.4; 2.5; 2.6; 4.5; 4.6; 4.8; 6; 7; 8. SIZE 4.5; 4.6; 4.8; 6; 7; 8; 17; 18.	*INSPECTED* LIMIT SIZE 4.10	*INSPECTED* LIMIT SIZE 4.8; 9; 11; 12;; 13; 17; 18.	*INSPEC.* LIMIT SIZE 4.8; 9; 10; 18.	*INSPEC.* LIMIT SIZE 4.7; 9; 18.

NOTE 1: The numbers in each column represent *thread gages* and *measuring equipment* specified in ASME B1.3M, Table 1. [10]
NOTE 2: ***LIMIT*** is inspected with a "thread limit gage"—***SIZE*** with a "thread indicating gage" or with other measuring equipment.
NOTE 3: Only System 23 requires inspection of surface texture.
NOTE 4: *This combination of limit and size inspection is only by agreement between purchaser and supplier (ASME B1.1-1989/¶9.1.2/pg. 53).[3]

Additional important aspects of each inspection standard are discussed in Notes 3 and 4 of Table II-2B of this report.

B. Comparing Systems 21, 22, And 23 In ASME B1.3M and in FED-STD-H28/20B

Systems 21-23 in FED-STD-H28/20B [14] accomplish essentially the same inspections as are accomplished by the inspections shown in Tables III-1A and III-1B of this report for Systems 21-23 in ASME B1.3M-1992 [10].

> System 21 in both standards requires inspection of the same thread elements and thread characteristics. The only thread element inspected by both System 21's for dimensional conformance or acceptability are: 1) major diameter for external thread, and 2) minor diameter for internal thread. Moreover, System 21 in both standards requires inspection of the thread characteristic called ***functional diameter*** by "**GO** Maximum Material" and "**NOT GO** Functional Diameter" inspections. (Gages for these inspections are made or set to the maximum (**GO**) and minimum (**NOT GO**) limits of size for pitch diameter specified in the applicable thread standard.)
>
> System 22 in both standards requires inspection of the same thread elements and thread characteristics. In the matter of dimensional conformance or acceptability, System 22 in both standards requires inspection of two-to-four thread elements for an external thread and two thread elements for internal threads. Furthermore, *System 22 in both standards requires inspection of pitch diameter.*
>
> System 23 in both standards requires inspection of the same thread elements and thread characteristics. In the matter of dimensional conformance or acceptability, System 23 in both standards inspects up to eleven thread elements for external threads and nine thread elements for internal threads. Moreover, *System 23 in both standards requires inspection of pitch diameter.*

In addition to the summaries of inspections specified in Systems 21, 22, and 23 which are listed in Tables III-1A and III-1B of this report, text in each inspection standard describes further aspects of the inspection systems. For example:

> ASME B1.3M-1992 (Section 6(c)) [10] *explicitly requires visual inspection* of the threaded product for gross defects such as missing or incomplete threads, defective thread profile, torn or ruptured surfaces and cracks, etc. On the other hand, FED-STD-H28/20B [14] *does not contain an explicit requirement for visual inspection* of the threaded product, but states (Section 5.2, Acceptability), that "Screw thread acceptability criteria are in accordance with Section 6 of ASME B1.3M-1992."
>
> Neither of these two standards explicitly specifies the *length of thread engagement*. This aspect of threaded-product inspection methodology is indirectly covered in each standard as explained in Notes 3 and 4 of Table II-2B of this report.

C. Degree of Inspection Severity

Systems 21, 22, and 23 in ASME B1.3M-1992 [10] and in FED-STD-H28/20B-1994 [14] offer a wide range of inspection severity. System 21 is least severe, System 22 is intermediate, and System 23 is most severe.

Tables III-1A and III-1B in this report, derived from Tables 1 and 3 in ASME B1.3M-92 [10], show the number of thread elements and thread characteristics examined by each inspection system. More elements and characteristics are inspected as inspection severity increases from System 21 to System 23.

The degree of inspection severity applied to threaded products is chosen by the designer/procurer of the product by the act of selecting an inspection method/gaging system specified in an inspection standard. The wide range of inspection severity is supplemented in the thread-inspection standard by a wide range of gages and measuring equipment from which the designer/procurer may select. Moreover, Sub-Paragraph 7(b) of ASME B1.3M-1992 [10] has provision to modify any of the three inspection systems so that thread elements or thread characteristics in addition to those already specified by System 21, System 22, or System 23, can be specified.

Mechanical reliability and durability required for successful application of the threaded product assembly are factors influencing the selection of inspection system. Guidance for selection of an inspection method or gaging system is offered in FED-STD-H28/20B [14] as follows:

> Consideration should be given to form, fit, function and fabrication of the threaded product. The costs of inspection should be compared to the possible costs resulting from uncertainties inherent in each inspection method.

IV. SCREW THREAD INSPECTIONS SPECIFIED IN MIL-S-7742 AND MIL-S-8879

A. Recent History

Table I-1 lists the history since 1959 of various editions of two military specifications which specify thread inspection, namely MIL-S-7742 and MIL-S-8879.

Prior to March 15, 1973, these two military specifications imposed rather severe inspection of threaded products. These documents explicitly specified that certain thread elements be within the required limits of size. For example:

> MIL-S-7742-B-1968: ¶4.1 *Inspection. Products having threads in accordance with this specification shall be inspected as stated herein and as specified on the detail standard, specification or production drawing.* NOTE: ¶s 3.5, 3.5.1 through 3.5.5, and 3.6 specify various aspects of gages and inspection procedures, e.g.,
>
> > 3.5.2 *Gaging Unified Class 3A threads. Class 3A threads require single element gaging in order to ascertain that all elements are within the required limits of size. Gages for external thread inspection shall be checked or set with setting plugs.*
>
> MIL-S-8879A-1965: ¶3.4 *Limits of Size. Screw threads, in accordance with this specification, shall be within the limits of size specified in tables II through VII, for the selected diameter-pitch combinations shown therein. The specified limits of size are considered exact and are inviolable unless specific exceptions are made.* NOTE: Tables II through VII specify maximum and minimum limits of size for such thread elements of external and internal threads as Major Diameter, Pitch Diameter, Minor Diameter, and Root Radius.

B. Introduction of Methods A, B, and C in 1973

Formally-specified screw thread inspection methods first appeared in MIL-S-7742 and MIL-S-8879 in 1973. For the first time, the military designer/procurer was offered a choice of inspection methods.

The March 15, 1973 amendments to MIL-S-7742B and MIL-S-8879C introduced three inspection methods, namely, Method A (for internal threads[7]), Method B (for external threads*), and Method C (for process control). The degree of inspection severity increased from Method A to Method C. Table IX from MIL-S-8879A — Amendment 1 (dated 15 March 1973, and Table III from MIL-S-7742B — Amendment 1 (dated 15 March 1973) describe the application of Methods A, B, and C.

---

[7] "Unless otherwise specified," as stated in ¶3.6 of MIL-S-7742B - AMENDMENT 1 and ¶3.5 of MIL-S-8879A - AMENDMENT 1. (15 March 1973)

Methods A, B, and C loosely correspond to Systems 21, 22, and 23, respectively. As Tables II-2A and II-2B of this report show, Methods A, B, and C predate by six years the first approval of Systems 21, 22, and 23 in a standardization document.

C. Effect of 1973 Amendments on Degree of Inspection Severity

One effect of the March 15, 1973 amendments was to offer the procurer/designer the opportunity to select an inspection method for certain threaded products which did not explicitly specify that certain thread elements be "within the required limits of size". In a sense, this option opened the door to less severe inspections than were specified in the pre-amendment editions of MIL-S-7742 and MIL-S-8879.

D. Changes in Inspection Methodologies in 1991

Inspections via Methods A, B, and C which were included prior to the 1988 limited Air Force editions of MIL-S-007742C and MIL-S-008879B, were completely eliminated in 1991. The 1991 editions of MIL-S-7742D and MIL-S-8879C permitted designation of the three inspection methodologies known as "System 21", "System 22", and "System 23."

In addition to the noted changes in Table II-1 of this report, the 1991 editions of these military specifications, MIL-S-7742D [8] and MIL-S-8879C [7], specified for the first time two thread-application categories. These categories, "Safety Critical Threads" and "Other Threads", are noted in Table II-1 of this report. One hundred percent inspection is required for "Safety Critical Threads".

V. GAGING SYSTEM ISSUES/CONCERNS

A. Can System 21 Assure Conformance to Dimensional Specifications?

1. System 21 Inspects Functional Diameter

ASME B1.3M-1992 [10] specifies System 21 as a screw-thread gaging system [¶4(b)(1), page 2].

> System 21 provides for inspection of the functional diameter of a threaded product, and for the major diameter of external threads, and for the minor diameter of internal threads. Inspections may be **limit** inspections or **size** inspections.
>
> System 21 does not provide for inspection of any other thread elements/characteristics.

System 21 inspection for functional diameter is intimately, but circuitously linked to the dimensional specifications for maximum pitch diameter and minimum pitch diameter which are tabulated in the applicable thread standard.

Two *goals* of System 21 inspection as specified in ASME B1.3M-1994 [10] and FED-STD-H28/20B [14] are to determine if the functional diameter of a threaded product is: 1) larger than the minimum pitch diameter, and 2) smaller than the maximum pitch diameter. In other words, System 21 inspection attempts to determine if the functional diameter is within the tolerances specified for pitch diameter. A third goal of System 21 inspection is to assure conformance to dimensional specifications for major diameter (external threads) and minor diameter (internal threads).

The result of System 21 inspection is determination of whether or not a threaded product will assemble with a mating threaded product which has passed similar inspection for functional diameter. Expectation is high that threaded products will assemble if the following condition is met:

(Minimum Pitch Diameter) < (**Functional Diameter**) < (Maximum Pitch Diameter)

2. Inspection Requirements of System 21

In ASME B1.3M-1992 [10], Tables 3 and 4 specify the thread characteristics and the thread element inspected by System 21. The *two thread characteristics* specified for inspection by System 21 are:

GO maximum material[8] and **NOT GO** functional diameter

---

[8] in ASME B1.3M-1992 - "**GO** maximum material" is the phrase used in Tables 3 and 4. "Maximum material **GO**" is the phrase used in Tables 1 and 2. These two phrases are used interchangeably throughout this report.

The *one thread element* specified for inspection by System 21 is:

Major Diameter for external product thread (Table 3), and
Minor Diameter for internal product thread (Table 4).

Table III-1A of this report lists some of the gages and measuring equipment specified in ASME B1.3M-1992 [10] for System 21 inspection of thread characteristics and thread elements. Table III-1A is constructed from information about *external product thread* inspection appearing in Tables 1 and 3 of ASME B1.3M-1992. Specified gages for **GO** maximum material and **NOT GO** functional diameter inspections include:

- Threaded Ring Gages, Split or Solid (**GO** or **NOT GO**)

- Thread Snap Gages with **GO** or **NOT GO** *segments*
- Thread Snap Gages with **GO** or **NOT GO** *rolls*

- Indicating Thread Gages — **GO** *segments* @ 120° or 180° contact
- Indicating Thread Gages — **GO** *rolls* @ 120° or 180° contact

NOTE: All specified gages are so designed and constructed that the length of engagement of product thread examined by System 21 "**GO** maximum material" and "**NOT GO** functional diameter" inspections encompasses several threads. Length of thread engagement during inspection is not explicitly mentioned for inspections specified in ASME B1.3M-1992 [10], but is included in the design information in ASME standards on gages and gaging, e.g., ANSI/ASME B1.2-1983 [11] (Appendix A, Table A4, p. 169), ASME B1.16M-1984 [15], and ASME B1.22M-85 [16]. As a general approximation, the length of thread engagement during inspection encompasses about 9-15 pitches or one diameter. Of course, **NOT GO** Functional Diameter ring and plug gages do not engage any threads if inspection is successful.

Table III-1A also lists the *functional limit inspections* and *functional size* inspections which are specified for System 21 in ASME B1.3M-1992 [10]. The procurer/designer and/or product supplier has the choice of selecting either *functional limit* or *functional size* inspections, although this point is not explicitly stated in ASME B1.3M-1992. Moreover, the procurer/designer and/or product supplier has the opportunity to specify for use in inspection any of the gages or measuring equipment listed in Table III-1A of this report or in ASME B1.3M-1992 (Tables 3 and 4).

3. Thread Characteristics Inspected by System 21

The two thread characteristics inspected by System 21 --- "maximum material **GO**" and "**NOT GO** functional diameter" --- are shown schematically in Figures V-1A and V-1B of this report, which represent schematically the System 21 inspection of an external product thread. Inspection for these two thread characteristics is linked to the maximum and minimum pitch diameters specified in the applicable thread standard.

The *upper bound magnitude* of the thread characteristic called "maximum material **GO**" is identical to the maximum pitch diameter (upper heavy dark line in Figures V-1A and V-1B) specified in the applicable thread standard.

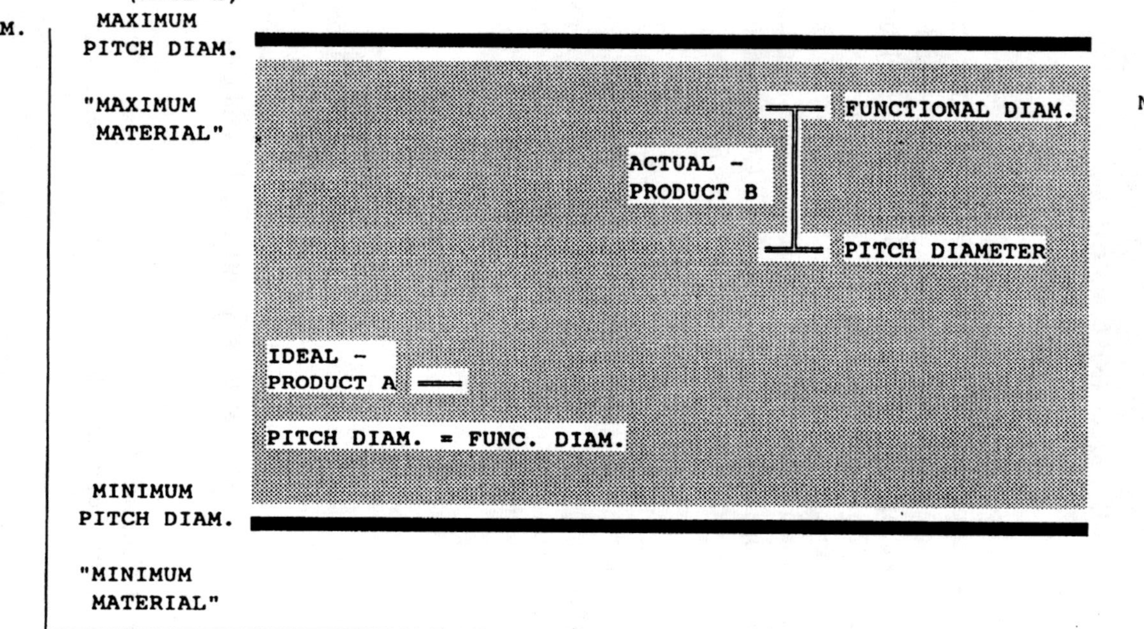

ACTUAL THREAD
(NOTE 3)

MAXIMUM MATERIAL GO	GO GAGES ARE MADE OR SET TO THE MAXIMUM PITCH DIAMETER SPECIFIED IN THE APPLICABLE THREAD STANDARD.

SUMMARY OF RESULTS FROM INSPECTING PRODUCTS A AND B* WITH SYSTEMS 21-22

SYSTEM 21	SYSTEM 22
ACCEPTABLE "MAX. MATERIAL GO"	ACCEPTABLE "MAX. MATERIAL GO"
ACCEPTABLE "NOT GO FUNCTIONAL DIAM."	MINIMUM MATERIAL—CONFORMING P.D.

NOT GO FUNCTIONAL DIAMETER	NOT GO GAGES ARE MADE OR SET TO THE MINIMUM PITCH DIAMETER SPECIFIED IN THE APPLICABLE THREAD STANDARD.

*PRODUCTS A AND B *WOULD* MEET THE REQUIREMENTS IN **MIL-S-8879C/¶3.4.5.1**.

NOTE 1. THE REFERENCE FEATURE FOR ALL THREAD INSPECTIONS IS THE *RANGE OF PITCH DIAMETER* SPECIFIED IN THE APPLICABLE THREAD STANDARD. MAXIMUM AND MINIMUM PITCH DIAMETER LIMITS OF SIZE, AS SPECIFIED IN ASME B1.1-1989 (e.g., TABLE 3A – pages 23-47) OR IN OTHER THREAD STANDARDS, ARE THE REFERENCE FEATURES FOR ALL GAGE MANUFACTURE/SETTING. ***GAGE TOLERANCES*** ARE SPECIFIED IN ANSI/ASME B1.2-1983, pages 18 AND 25.

NOTE 2. AN IDEAL THREAD HAS A ***PITCH DIAMETER SIZE*** WHICH FALLS WITHIN THE RANGE OF THE MAXIMUM AND MINIMUM PITCH DIAMETERS. AN *IDEAL THREAD* HAS NO VARIATIONS IN THREAD ELEMENTS/CHARACTERISTICS. ALL SIZES AND GEOMETRIC PROFILES ARE PERFECT.

NOTE 3. AN *ACTUAL PRODUCT THREAD* HAS VARIATIONS IN THREAD ELEMENTS/CHARACTERISTICS – NOTABLY, IN PITCH, LEAD, FLANK ANGLE, TAPER, STRAIGHTNESS AND CIRCULARITY (ROUNDNESS). AN ACTUAL PRODUCT THREAD HAS A ***FUNCTIONAL DIAMETER SIZE*** WHICH IS DETERMINED BY THE MAGNITUDES OF VARIATIONS IN THREAD ELEMENTS/CHARACTERISTICS.
THE FUNCTIONAL DIAMETER SIZE INCORPORATES A " ... CUMULATIVE THREAD ELEMENT VARIATION DIFFERENTIAL".
THE DIFFERENCE BETWEEN FUNCTIONAL DIAMETER AND PITCH DIAMETER " ... REPRESENTS THE DIAMETRAL EFFECT OF THE TOTAL AMOUNT OF THREAD ELEMENT VARIATIONS." (ANSI/ASME B1.2-1983 – page 37)
EXTERNAL THREAD: THE FUNCTIONAL DIAMETER SIZE IS ALWAYS LARGER THAN THE PITCH DIAMETER SIZE.
INTERNAL THREAD: THE FUNCTIONAL DIAMETER SIZE IS ALWAYS SMALLER THAN THE PITCH DIAMETER SIZE.

FIGURE V-1A. COMPARISON OF INSPECTION CONSIDERATIONS (**NOTE 1**) FOR PITCH DIAMETER AND FUNCTIONAL DIAMETER – IDEAL THREAD (PRODUCT A) AND ACTUAL THREAD (PRODUCT B) – **EXTERNAL PRODUCT THREAD**

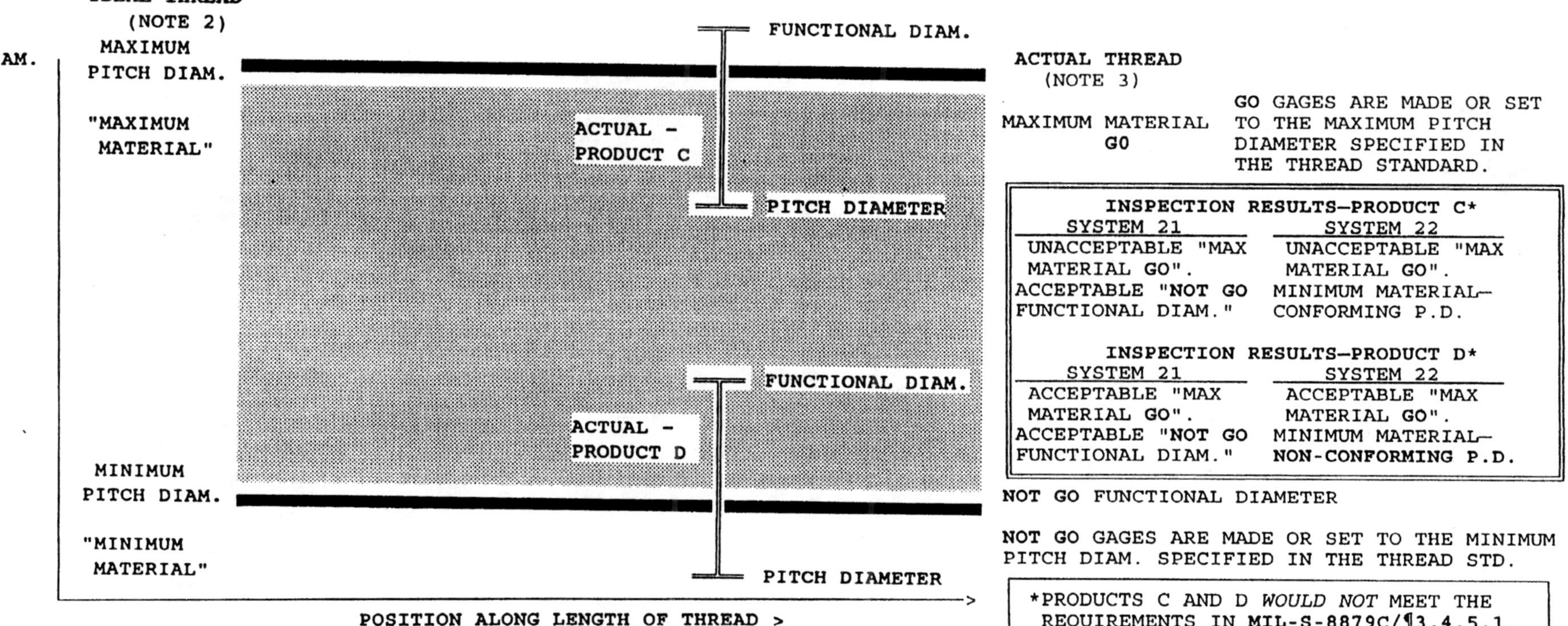

NOTE 1. THE REFERENCE FEATURE FOR ALL THREAD INSPECTIONS IS THE *RANGE OF PITCH DIAMETER* SPECIFIED IN THE APPLICABLE THREAD STANDARD. MAXIMUM AND MINIMUM PITCH DIAMETER LIMITS OF SIZE, AS SPECIFIED IN ASME B1.1-1989 (e.g., TABLE 3A – pages 23-47) OR IN OTHER THREAD STANDARDS, ARE THE REFERENCE FEATURES FOR ALL GAGE MANUFACTURE/SETTING. ***GAGE TOLERANCES*** ARE SPECIFIED IN ANSI/ASME B1.2-1983, pages 18 AND 25.

NOTE 2. AN IDEAL THREAD HAS A ***PITCH DIAMETER SIZE*** WHICH FALLS WITHIN THE RANGE OF THE MAXIMUM AND MINIMUM PITCH DIAMETERS. AN *IDEAL THREAD* HAS NO VARIATIONS IN THREAD ELEMENTS/CHARACTERISTICS.

NOTE 3. AN *ACTUAL PRODUCT THREAD* HAS VARIATIONS IN THREAD ELEMENTS/CHARACTERISTICS – NOTABLY, IN LEAD, FLANK ANGLE, TAPER, STRAIGHTNESS, AND CIRCULARITY (ROUNDNESS). AN ACTUAL PRODUCT THREAD HAS A ***FUNCTIONAL DIAMETER SIZE*** WHICH IS DETERMINED BY THE MAGNITUDES OF VARIATIONS IN THREAD ELEMENTS/CHARACTERISTICS.
THE FUNCTIONAL DIAMETER SIZE INCORPORATES A " ... CUMULATIVE THREAD ELEMENT VARIATION DIFFERENTIAL".
THE DIFFERENCE BETWEEN FUNCTIONAL DIAMETER AND PITCH DIAMETER " ... REPRESENTS THE DIAMETRAL EFFECT OF THE TOTAL AMOUNT OF THREAD ELEMENT VARIATIONS." (ANSI/ASME B1.2-1983 – page 37)
EXTERNAL THREAD: THE FUNCTIONAL DIAMETER SIZE IS ALWAYS LARGER THAN THE PITCH DIAMETER SIZE.
INTERNAL THREAD: THE FUNCTIONAL DIAMETER SIZE IS ALWAYS SMALLER THAN THE PITCH DIAMETER SIZE.

FIGURE V-1B COMPARISON OF INSPECTION CONSIDERATIONS (**NOTE 1**) FOR PITCH DIAMETER AND FUNCTIONAL DIAMETER – ACTUAL THREADS (PRODUCTS C AND D) – **EXTERNAL PRODUCT THREAD**

The *lower bound magnitude* of the thread characteristic called "**NOT GO** functional diameter" is identical to the minimum pitch diameter (lower heavy dark line in Figures V-1A and V-1B) specified in the applicable thread standard.

The product *thread itself* is characterized by two related diameters:

Functional Diameter — a thread characteristic linked to the ability of a product thread to assemble. The functional diameter of an external product thread is usually larger than its pitch diameter.

Pitch Diameter — a thread element. Maximum and minimum pitch diameters are listed in the applicable thread standard. The pitch diameter is used as the reference location in a perfect thread profile to dimension the tolerance envelope.

Four hypothetical threaded products (external threads) are represented schematically in Figures V-1A and V-1B. They are labeled Products A-D. Each product has a functional diameter and a pitch diameter. The functional and pitch diameters are depicted as the labeled horizontal bars in each figure.

Ideal Product A in Figure V-1A has a functional diameter equal to its pitch diameter. *Ideal Product A exists in imagination only.*

Actual Product B in Figure V-1A and actual Products C and D in Figure V-1B, have functional diameters which are larger than their pitch diameters. Functional diameter is determined by the magnitudes of variations in thread elements and thread characteristics and includes the cumulative effect of these variations ("equivalent change in functional diameter"[9]) added to pitch diameter.

4. Objectives of System 21 Inspection

One objective of inspection by System 21 is to assure that the functional diameter of the external product thread does not exceed the maximum pitch diameter (upper bound of "**GO** maximum material").

Inspection by System 21 also seeks to assure that the functional diameter of the external product thread does not fall below the minimum pitch diameter (lower bound of "**NOT GO** functional diameter").

When these two objectives are met, the external product thread can be counted upon to assemble with a mating threaded product which has passed similar inspection for functional diameter. In fact, *Thread Assembly Diameter* is an alternative term for "functional diameter" given in ANSI/ASME B1.7M-1984 (page 6) [1].

5. System 21 Does Not Require Inspection for "Minimum Material"

Informal discussions of System 21 thread inspection relative to the lower bound of "**NOT GO** functional diameter" often imply that this inspection procedure "determines whether or not *the minimum material condition* is met." Furthermore, FED-STD-H28/20B [14] refers to System 21 gages (¶5.1.3.1) and states,

[9] "Equivalent change in functional diameter" is the phrase used in the heading of Table 4 in ASME B1.1-1989 (page 55) to describe the allowable variation in lead.

"These functional gages *provide some control at the minimum material limit* when there is little variation in thread form characteristics such as lead, flank angle, taper and roundness." Such viewpoints are misleading because examination for *meeting* the minimum material requirement is not intended by System 21 inspection. However, examination for *not meeting* the minimum material requirement can be accomplished by **NOT GO** Functional Diameter inspection. For example, if an external product thread successfully enters a **NOT GO** Functional Diameter ring gage, it is clear that both the functional diameter and the pitch diameter of the external product thread are less than the minimum pitch diameter specified in the applicable thread standard.

> The three minimum material inspections specified in ASME B1.3M-1992 [10] are set forth in the first three columns of Tables III-1A and III-1B of this report. These inspections are specified only for System 22 and System 23.
>
> Minimum material inspection of a threaded product aims to determine whether or not its pitch diameter is larger or smaller than the minimum pitch diameter (lower heavy dark line in Figures V-1A and V-1B) specified in the applicable thread standard.

Determination of whether or not the pitch diameter of an external product thread is below the minimum pitch diameter may not be possible using the gages specified for System 21 inspection of "**NOT GO** functional diameter" in ASME B1.3M-1992 [10] or in FED-STD-H28/20B [14]. The reason is that System 21 gages inspect by engaging threads over several pitches whereas pitch diameter can be measured reliably only by using gages which inspect over a distance of one pitch. The essential technical point is that variations in the longitudinal character of a product thread can affect the outcome of inspections made with gages specified with System 21, whereas variations in the longitudinal character of a product thread cannot affect the outcome of inspections made with gages specified for pitch diameter inspection.

> The flank engagement between two and one-half external product threads and two and one-half threads of a **NOT GO** functional diameter Thread Ring Gage is shown in Figure V-2 of this report, which is taken from ANSI/ASME B1.2-1983 (Figure 19, page 129) [11]. The external product threads and the threads of the Ring gage shown in Figure V-2 are of perfect thread form. The external product threads are of minimum pitch diameter.
>
> The contact between a single external product thread and the gaging elements of a Thread Snap Gage — minimum-material pitch diameter limit — cone and vee — is shown in Figure V-3 of this report, which is taken from ANSI/ASME B1.2-1983 (Figure 22, page 134) [11].

6. System 21 and Conformance to Dimensional Specifications for Functional Diameter and Pitch Diameter

In terms of System 21 capabilities for determining conformance to dimensional specifications for functional diameter[10] and pitch diameter limits specified in the applicable thread standard, the following tabulation lists results from System 21 inspections of hypothetical Products A-D in Figures V-1A and V-1B of this report.

---

[10] *Functional diameter* does not appear as a dimensional specification in any thread standard. An implicit assumption in screw thread inspection technology is that *functional diameter limits* are identical to *pitch diameter limits*, i.e.,that *maximum functional diameter limit* equals maximum pitch diameter, and that *minimum functional diameter limit* equals minimum pitch diameter.

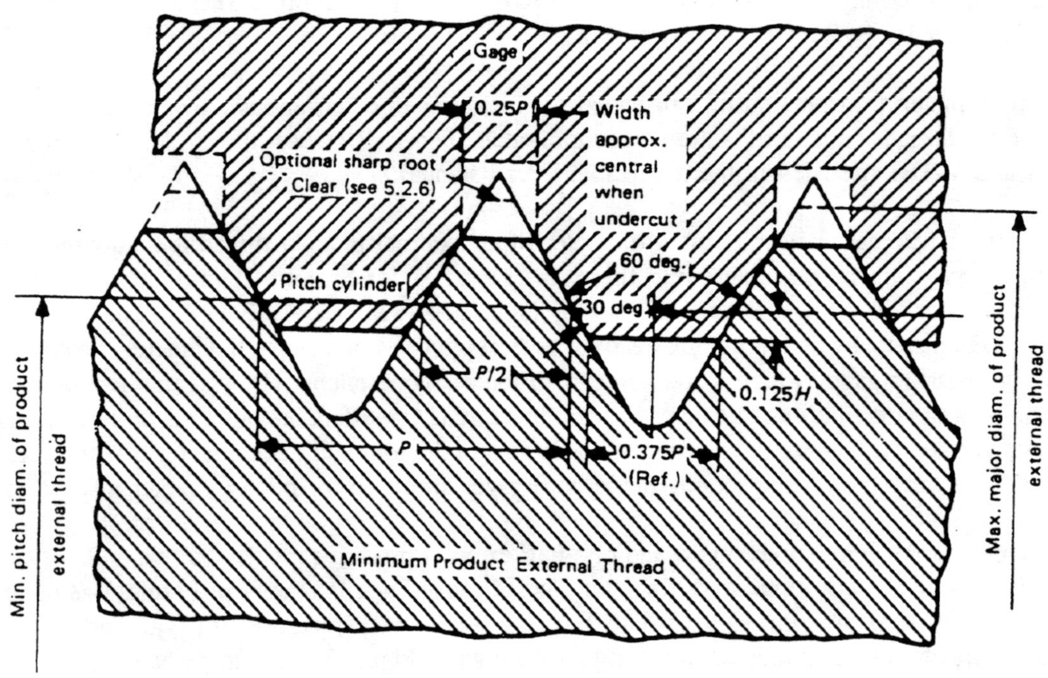

V-2. Flank engagement between two and one-half perfect external product threads and two and one-half threads of a **NOT GO** functional diameter Thread Ring Gage.

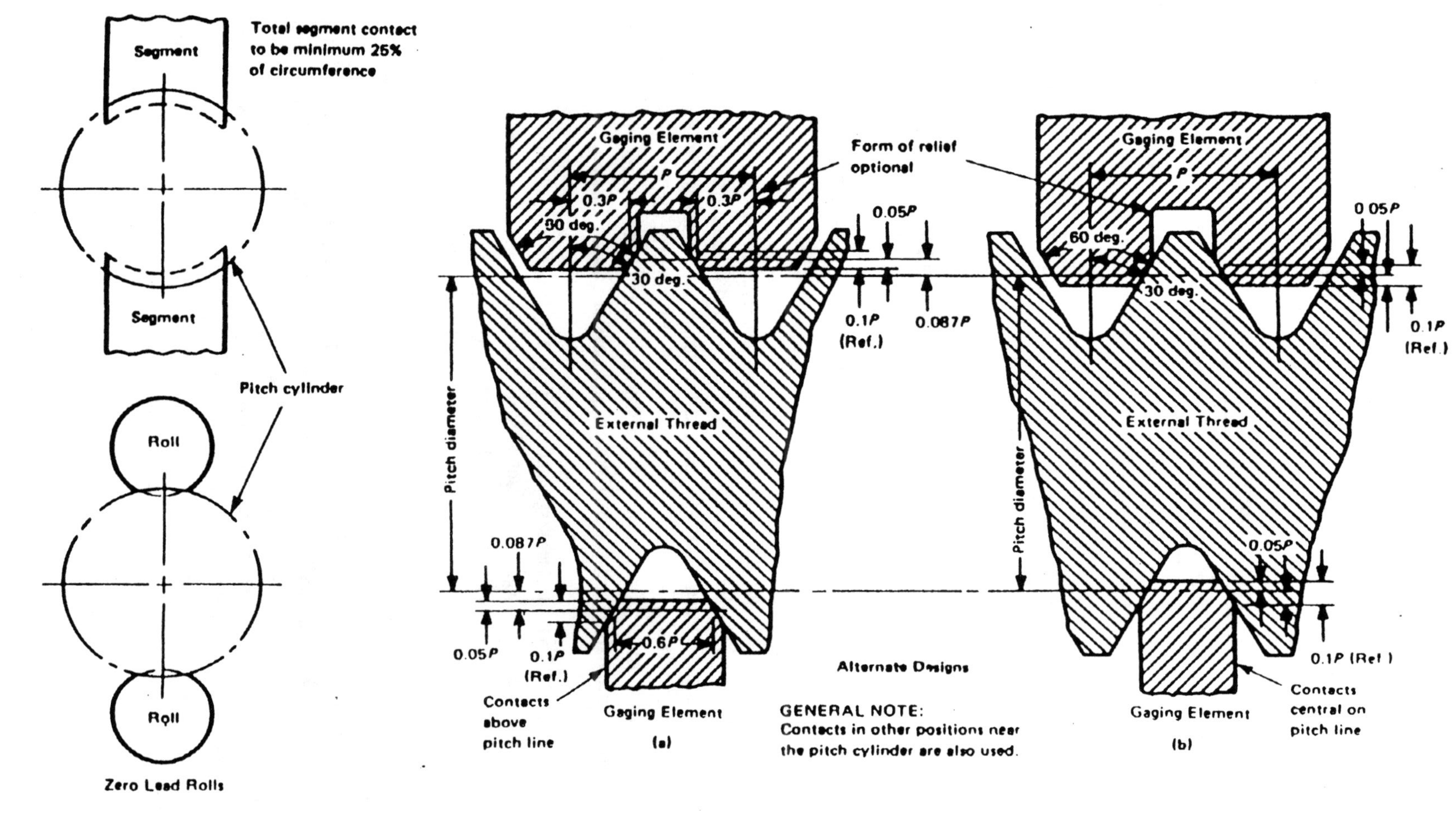

V-3 Contact between a single external product thread and the gaging elements of a Thread Snap Gage — minimum-material pitch diameter limit — cone and vee.

PRODUCT	FUNCTIONAL DIAMETER*	PITCH DIAMETER
A	Ideal Thread at Maximum Pitch Diameter Limit	At Maximum Pitch Diameter Limit
B	Within Pitch Diameter Limits	Within Pitch Diameter Limits
C	Larger Than Max. Pitch Diameter	Within Pitch Diameter Limits
D	Within Pitch Diameter Limits	Less Than Min. Pitch Diam.

*NOTE: *Functional diameter* does not appear as a dimensional specification in any thread standard. An assumption in thread inspection technology is that *functional diameter limits* are identical to *pitch diameter limits*, i.e., that *maximum functional diameter limit* equals maximum pitch diameter, and *minimum functional diameter limit* equals minimum pitch diameter.

In terms of inspection for functional diameter and pitch diameter by System 21, it is evident from the above tabulation that:

Product A (Figure V-1A) can be inspected reliably for pitch diameter with the gages specified in ASME B1.3M-1992.

Product B (Figure V-1A) can be inspected reliably for functional diameter with the gages specified in ASME B1.3M-1992. Furthermore, the pitch diameter of Product B fortuitously conforms to dimensional specifications for pitch diameter, even though pitch diameter inspection is not explicitly intended via System 21 inspection. Conformance of Product B to functional diameter and pitch diameter dimensional specifications is assured. Moreover, Product B will assemble with a mating threaded product which has passed similar inspections.

Product C (Figure V-1B) can be inspected reliably for functional diameter with the gages specified in ASME B1.3M-1992. Product C will not pass inspection because its functional diameter is larger than the maximum pitch diameter. An equivalent way of saying this is that Product C will be rejected because its "**GO** maximum material" thread characteristic exceeds the specified maximum pitch diameter. This result means that the question of whether or not Product C can assemble satisfactorily is moot.

Product D (Figure V-1B) can be inspected reliably for functional diameter with the gages specified in ASME B1.3M-1992. The specified gages can determine whether or not the functional diameter of the external product thread is larger or smaller than the minimum pitch diameter, but none of these gages can determine that the pitch diameter of Product D is smaller than the specified minimum pitch diameter. Consequently, System 21 inspection of Product D can assure its satisfactory assembly, but System 21 inspection cannot assure that Product D will conform to dimensional specifications for minimum pitch diameter.

7. System 21 and Conformance to Dimensional Specifications

Table III-1A of this report shows that System 21 requires inspection for one thread element and two thread characteristics.

Inspection of the specified thread element (major diameter/external thread or minor diameter/internal thread) can be carried out reliably according to System 21 to check conformance of major or minor diameters to dimensional specifications.

Inspection of the thread characteristic, "**GO** maximum material", can be carried out reliably according to System 21 to check conformance of functional diameter to dimensional specifications for the upper bound of "**GO** maximum material" (upper heavy dark line in Figures V-1A and V-1B, which is identical to maximum pitch diameter).

Inspection of the thread characteristic, "**NOT GO** functional diameter", can be carried out reliably according to System 21 to check conformance of functional diameter to dimensional specifications for the lower bound of "**NOT GO** functional diameter" (lower heavy dark line in Figures V-1A and V-1B, which is identical to minimum pitch diameter).

NOTE: Inspection of the thread characteristic, "**NOT GO** functional diameter" with the gages specified for "**NOT GO** functional diameter" inspection may not assure conformance but assure nonconformance of pitch diameter of a threaded product to dimensional specifications for pitch diameter. Nor can this inspection determine if the minimum material is *met*. However, "**NOT GO** functional diameter" inspection can determine if the minimum material condition is *not met*.

System 21 does not require inspection for any other thread elements or thread characteristics, as is evident from examination of Tables III-1A and III-1B of this report. For example, System 21 is not used to evaluate thread elements such as lead or flank angle, or for thread characteristics such as circularity (roundness) or axial taper, and consequently, System 21 cannot assure conformance to dimensional specifications for those thread elements/characteristics. Nevertheless, some control is provided by the **GO** and **NOT GO** functional gages.

B. Can Systems 22 and/or 23 Assure Conformance to Dimensional Specifications?

1. System 22 Inspection

ASME B1.3M-1992 (¶4(b)(3), page 2) [10] specifies System 22 as a screw-thread gaging system.

Minimum Material: System 22 specifies inspection for "minimum material" of a threaded product via pitch diameter or thread groove diameter inspection, as shown in Table III-1B of this report. Table III-1B also lists a third inspection for "minimum material" which is a combination of limit and size inspections carried out only by agreement between purchaser and supplier.[11]

Other Inspections: System 22 also specifies inspection for the thread characteristic called "maximum material **GO**", and for the thread elements: 1) major or minor diameter, and 2) root profile.

Inspection using System 22 can check conformance to dimensional specifications for thread characteristics and thread elements. As Table III-1B shows, System 22 does not specify inspection for:

---

[11] See ¶9.1.2 on page 53 in ASME B1.1-1989 (3).

1) roundness of pitch cylinder

2) taper of pitch cylinder

3) cumulative form variation

4) lead including helix variation

5) flank angle variation

6) runout major diameter to pitch diameter

Consequently, inspection by System 22 cannot assure conformance to dimensional specifications for these thread characteristics and thread elements.

2. System 23 Inspection

ASME B1.3M-1992 [10] specifies System 23 as a screw thread gaging system in ¶4(b)[4] on page 2 of the standard.

Minimum Material: System 23 specifies inspection for "minimum material" of a threaded product via pitch diameter or thread groove diameter inspection, as shown in Table III-1B of this report.

Other Inspections: System 23 also specifies inspection for the thread characteristic called "maximum material **GO**", and for the thread elements: 1) major or minor diameter, and 2) root profile. Moreover, System 23 specifies inspection for:

1) roundness of pitch cylinder

2) taper of pitch cylinder

3) cumulative form variation

4) lead including helix variation

5) flank angle variation

6) runout major diameter to pitch diameter

Consequently, inspection by System 23 can assure conformance to dimensional specifications for these thread characteristics and thread elements.

C. Engineering Data Relating Performance of Threaded Assembly with Deviation from Dimensional Specifications

1. Assembly and Mechanical Performance

ASME B1.3M-1992 (10) specifies inspection systems which result in "dimensionally acceptable" products, as might be expected from the title of this standard, *Screw Thread Gaging Systems for Dimensional Acceptability — Inch and Metric Screw-Threads.* Successful inspection with gaging systems specified in this standard delivers threaded products which should assemble with mating products. This point is reinforced by an Industrial Fasteners Institute document which references ASME B1.3M-1992 [10], namely, "Recommendations For Fastener Thread Acceptability." [17]

Will the threaded assembly stay together and perform its intended design function?

This question is left in the hands of the designer/procurer of threaded products. ASME B1.3M-1992 [10] does not address this issue of the mechanical performance of the threaded assembly.

Features of mechanical performance considered by designers include: axial and torsion strength, shear strength, resistance to vibration loosening, fatigue resistance, and hydraulic pressure integrity.

Conventional engineering wisdom has focused on the *pitch diameter* of threaded products as one of the most important thread elements determining mechanical performance of a threaded assembly. (In contrast, *functional diameter* usually is not a primary consideration in discussions of mechanical performance.)

One view, supported by long experience, is that satisfactory mechanical performance results if pitch diameters of the threaded products in an assembly are within the maximum and minimum limits of pitch diameters specified in the applicable thread standard, and if the form of thread satisfies dimensional acceptability criteria specified in the applicable screw thread gaging standard.

Another view, supported by results of calculations [18, 19], is that satisfactory mechanical performance may result even if pitch diameters of the threaded products in an assembly do not conform to dimensional specifications for pitch diameter noted in the applicable thread standard. Equations for these calculations appear in ASME B1.1-1989 (Appendix B, page 141) [3], and in FED-STD- H28/2B [20].

Conventional engineering wisdom also has considered the other thread elements (besides pitch diameter) and thread characteristics listed in Table III-1B and has concluded that if thread elements and thread characteristics of threaded products conform to dimensional specifications for numerous thread elements and thread characteristics, an assembly of threaded products will exhibit satisfactory mechanical performance. This conclusion is supported by long experience. However, technical documentation in support of this conclusion is scarce.

It is generally believed that a threaded assembly is held together by interference and friction between the internal and external threads. Consequently, flank engagement between internal and external threads is an important factor influencing the mechanical performance of a threaded assembly in service.

> Interference between internal and external threads results from slight differences in dimensions of the thread elements and thread characteristics of the internal and external thread.

Interference between internal and external threads can also result from elastic deformation or plastic deformation of the threads during assembly.

Conventional engineering wisdom [21-24] asserts that threaded products which have passed inspection specified by a screw thread gaging system will perform their intended design function provided the threaded assembly is put together with appropriate tensile load. Presumably, an appropriate torque results in sufficient preload to hold the threaded assembly together throughout the service life of the assembly.

In practice, conservative engineers take further precautions to assure that a threaded assembly stays together in service. These precautions include:

1) periodic inspection and re-torquing of the threaded assembly.

2) use of mechanical locking devices, e.g., lock wires, lock nuts, lock washers.

3) polymeric films or polymeric inserts in the joint.

4) designs which assemble a tapered thread with a straight thread (if controlled preload not required).

5) use of load indicating-devices which can show that preload tension is satisfactory.

These precautions can help to overcome service conditions which might keep a threaded assembly from staying tightly together. Such service conditions include: vibration; large, cyclic mechanical loads; dimensional variations arising from thermal cycling; and galling.

2. Inspection Results Compared with Dimensional Specifications

Some technical data is available in the public domain which indicates that some externally threaded products held in inventory have pitch diameters smaller than the minimum pitch diameter specified in the applicable thread standard [25, 26].

Such a result is not unexpected for product inventory inspected by System 21, because the design of the **NOT GO** gaging specified by System 21 in ASME B1.3M-1992 [10] and FED-STD-H28/20 [14] may not detect product pitch diameter which is not in conformance with dimensional specifications. This point is discussed in Section V-A-6 of this report. Moreover, this point has been addressed in screw thread standardization documents since at least 1957, as shown in the discussion on pages 108-109 of NBS Handbook H28 (1957) [5b], which includes the following statements:

> Also, it is a principle of limit gaging that each element or dimension can be checked only singly by a minimum-metal-limit gage.
>
> It is not feasible, therefore, to establish an ideal gage design for gaging pitch diameter and approach that ideal closely in practice, as is done for maximum-metal-limit gages.

A comprehensive screw thread inspection using standardized gaging systems was conducted in 1971-1972 by the National Bureau of Standards in cooperation with the Industrial Fastener Institute [27]. A study was made of six screw thread/pitch diameter combinations of 10 samples each, both external and internal threads. Thread elements inspected included: pitch, flank angle, major diameter, minor diameter, and thread-groove pitch diameter. Findings of this study are not in the public domain.

3. Calculations of Joint Strength Linked to Pitch Diameter Variations

Aerospace Industries Association [18]: Calculated results according to FED-STD-H28/2B led to the conclusion that "thread shear (stripping) failures are not critical for structural integrity or flight safety when bolt/nut joints are designed by good aircraft design practices and are within two times MIL-S-8879C [Ref. 7 of this report] tolerances." This study based on calculations using equations in FED-STD-H28/2B [20].

Study By H.W. Ellison [19]: Calculated results according to a proprietary model [19] led to the conclusion that "Threads with lead and pitch diameter outside intended specifications may be found acceptable when inspected using attribute gages. Threads with these errors do not affect the strength of the part."

Studies By E.M. Alexander [27a, 27b]: Calculated results using a model developed by Alexander for bolt-stripping loads and thread-stripping loads for "inch series fasteners" led to the conclusion that "as nut height increases the probability of stripping decreases and the probability of bolt breaking shows a complimentary increase." The focus of these two papers by Alexander is on mechanical performance, which is affected by: 1) overall sizes of threaded products and length of thread engagement, 2) materials, and 3) torque. These papers by Alexander are often referred to in discussions of how variations in thread characteristics/elements affect mechanical performance. However, the effects of size variations of thread elements and thread characteristics on mechanical performance of threaded assemblies are not addressed by Alexander.

4. Laboratory Measurements This survey of interested parties with concerns about screw thread gaging systems, and particularly with concerns about how size variations of thread elements and thread characteristics affect mechanical performance of threaded assemblies, led to the following findings about technical data:

The Industrial Fasteners Institute is finalizing a laboratory study in which tests were conducted for both commercial and aerospace products to determine the relationship of pitch diameter variations in bolts and nuts to the mechanical and performance characteristics of threaded fasteners in assembly. The preliminary results indicate that there is very little direct correlation between pitch diameter size variation of fasteners and their resulting performance. It is expected that this paper will be published in the public domain in 1996 [29].

The Johnson Gage Company is currently finalizing the testing stage of a laboratory study in which tests were conducted to determine how variations in pitch diameter of threaded products affects static tensile strength, torsional strength, and vibration loosening of threaded assemblies [30]. Findings of this study are not in the public domain.

A study [31] titled, *Test Data of Internal Threads*, is referred to in Reference 18 of this study. However, findings are not in the public domain.

5. Text Books

Two text books provide some information about the functional performance of threaded assemblies [21, 22]. However, these books do not offer substantive technical data relating functional performance of threaded products with the degree to which thread elements or thread characteristics of a threaded product deviates from specified dimensions.

6. Failure Analysis Report

UH-60 AIRCRAFT/SPINDLE & SPINDLE NUT
ARMY SAFETY CENTER/CORPUS CHRISTI ARMY DEPOT/REPORT NO 85MX090 04/11/85

Some technical data are available in the public domain which suggests that "helical path deviation of the thread" contributed to failure via fatigue mechanisms [25]. This interpretation is contested between the provider and the customer. More information about this incident is available via a Freedom Of Information request to:

Public Information Officer **AMSAT-B-P,** U.S. Army Aviation and Troop Command
4300 Goodfellow Boulevard, St. Louis, MO 61320 **(314) 263-11674**

This is the only documented failure linked to a threaded product which has appeared in the public domain during this study.

D. Concerns Raised by National Screw-Thread Community

1. Makeup of the National Screw-Thread Community

The national screw-thread community consists of 1) the custodians and caretakers of screw-thread standardization documents, 2) the metrologists who make, calibrate, and otherwise maintain gages and measuring equipment for screw-threads, 3) the manufacturers of screw-thread products, and 4) the users of screw-thread products.

2. Soliciting Statements

ASME/CRTD convened a panel (as described in Section IA of this report) which instituted several administrative procedures to identify and investigate issues/concerns about threaded products linked to inspection standards and specifications, metrology, and mechanical performance. Two activities were carried out to facilitate the identification of issues of concern to the national screw-thread community.

A meeting of interested parties was convened on August 12, 1994, in which all parties were invited to 1) identify issues/concerns linked to screw-thread inspection technology, 2) provide technical documentation about the issues/concerns, and 3) suggest options for resolutions. A form for stating issues/concerns was made available to interested parties to complete and submit to ASME/CRTD.

NIST and ASME/CRTD issued a joint press release about this study in the fall of 1994. The press release invited interested parties to contact the ASME/CRTD office in Washington, DC or ASME/CRTD's Principal Investigator. Interested parties were invited to: 1) identify issues/concerns linked to screw-thread inspection technology, 2) submit technical documentation about issues/concerns, and 3) suggest options for resolving the issues/concerns.

3. Issues Identified

Seven issues were identified and are listed below.

I. Extent to which each gaging system can assure conformance to dimensional specifications.

II. References to incidents resulting from alleged non-conforming threads and mechanical performance.

III. Unclear meaning of language/terminology used in certain standards, e.g., "functional diameter"; "minimum material".

IV. Need for information cross-links between/among standardization documents coupled to other standards.

V. Uncertainty about how error propagation in a sequence of calibrations, (i.e., calibration of master gages using gage blocks and calibration of working gages using master gages) affects reliability and repeatability of the calibration of the working gages.

VI. Shortage of technical data about mechanical performance of threaded assemblies.

VII. Excessive costs linked to meeting some requirements of certain military specifications.

Little supporting documentation accompanied the statements about issues and/or concerns. Information provided by interested parties about issues/concerns is mainly in the following categories: 1) policy statements, 2) sales literature for thread gages, 3) anecdotes and incomplete documentation about non-conformities of threaded products, and 4) text-books about threaded products.

The U.S. Nuclear Regulatory Commission provided what appears to be the most thorough technical documentation in the public domain about inspections of inventories of threaded products [26]. These inspections were carried out to determine conformance to dimensional specifications.

Several standardization documents were obtained from the custodians of the documents: American Society of Mechanical Engineers; Defense Industrial Supply Center; and U.S. Air Force.

Issue I — Gages - One long standing concern is linked to a question about the ability of a **NOT GO** threaded ring or plug gage to inspect for "minimum material". FED-STD-H28/20B [14] suggests that these **NOT GO** gages can make such an inspection, though not fully, whereas ASME B1.3M-1992 [10] does not address this concern. NBS (NIST) has stated repeatedly since 1977 that these **NOT GO** gages cannot make such an inspection [32].

Table III-1A of this report, constructed from information in ASME B1.3M-1992 (10), demonstrates that **NOT GO** gages are not specified to inspect for minimum material. Furthermore, **NOT GO** gages are not designed to inspect for minimum material, as discussed in Section V-A-5 of this report.

A second concern about gages was raised which involves the thread form of the **NOT GO** threaded ring gage, shown in Figure V-2 of this report, which is taken from ANSI/ASME B1.2-1983 (Figure 19, page 129) [11]. The concern is about the small amount of flank engagement for the **NOT GO** gage, as compared to the larger amount of flank engagement for the **GO** threaded ring gage shown in Figure V-4 of this report, which is taken from ANSI/ASME B1.2-1983 (Figure 18, page 127) [11]. As these two figures show, the **NOT GO** threaded ring gage (Figure V-2) is designed for less flank engagement than the **GO** threaded ring gage (Figure V-4), since the radial distance a thread projects beyond the pitch cylinder for the **NOT GO** gage is specified as 0.125H, whereas the radial distance a thread projects beyond the pitch cylinder is specified as 0.250H for the **GO** gage.

Of interest in connection with this concern are two points:

* Interference is more likely between the threaded ring gage and the threaded product inspected as flank engagement increases.

* The wear-life of the threaded ring gage is likely to be shorter as the amount of flank engagement decreases.

A third concern about gages involves variations in results of thread inspection using indicating gages from different manufacturers [33]. This concern suggests that some attention be focused on critical evaluation of the accuracy of indicating gages. One suggestion is that this concern arises from the design of segments and rolls in the indicating gages.

Issue II — Alleged Non-Conforming Threads Causing Incidents: A concern much on the minds of everyone in the national screw-thread community is the many references and dramatic overstatements about incidents linked to alleged non-conforming threads and resulting faulty mechanical performance. However, very little persuasive substantiating technical data or case studies have been provided in support of these allegations.

What is needed to evaluate the concern about non-conforming threads is thorough and traceable technical documentation in the public domain about results of inspections of threaded products, and comparison of these results with requirements of the gaging system according to which the threaded products were procured.

What is needed to evaluate the allegations of incidents being caused by non-conforming threads is thorough and traceable technical documentation in the public domain about the alleged incidents. Persuasive documentation would be in the form of complete failure analysis reports from an established, reputable, and disinterested institution.

Any documentation which exists with regard to non-conforming threads causing incidents is not likely to be made available to this study because of product liability litigation concerns.

Issue III — Unclear Meanings in Certain Standards: A concern has been raised about unclear language and terminology used in certain standards.

Examples of terminology which is regarded as having unclear meanings are: "functional diameter", "minimum material", and "attributes". A GLOSSARY has been created as part of this study of screw-

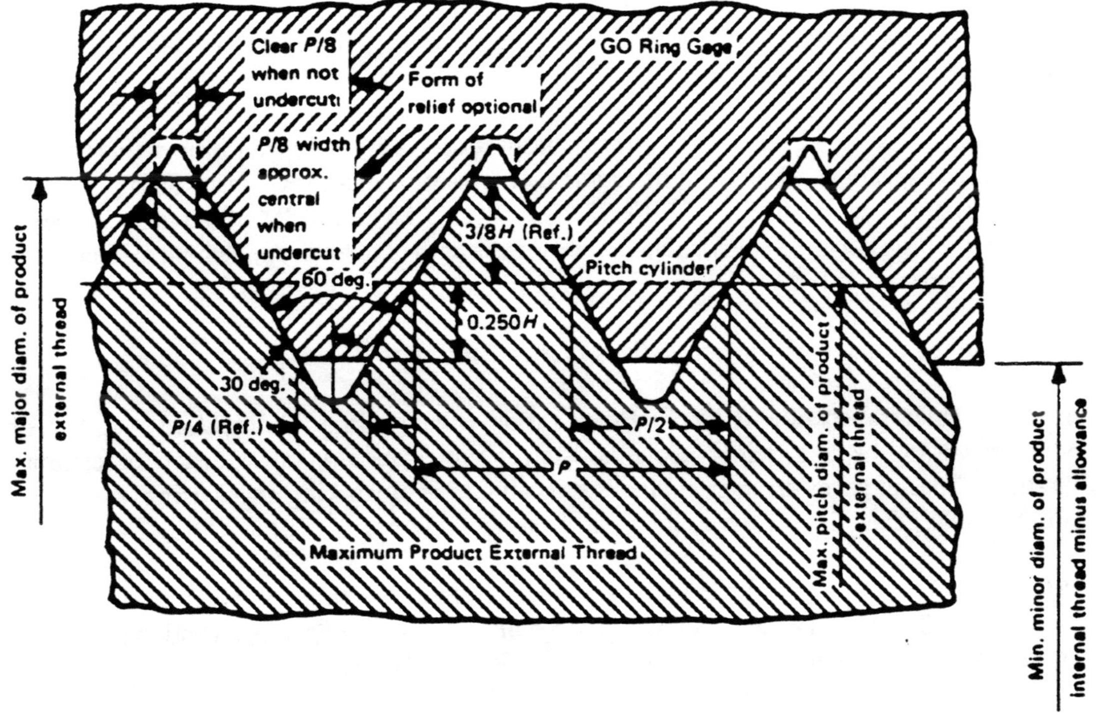

V-4 Design features of a GO Thread Ring Gage. The radial distance a thread projects beyond the pitch cylinder is 0.250H [11].

thread gaging systems to clarify the terms used throughout the reports being written for this study. The GLOSSARY is included in this Report as an Appendix B.

Paragraph 4(b)-(1)[12] in ASME B1.3M-1992 [10] is an example of language which has unclear meaning. In fact, this unclear meaning is probably related to the widespread confusion about the goals and results of System 21 inspection. Concerns about unclear language can best be dealt with by the custodians of the standardization documents, who admittedly have difficulty writing a clear consensus standard on a controversial subject.

Issue IV — Information Cross-Links Between/Among Standardization Documents: A concern was raised about the need for information cross-links between/among standardization documents. Examples and illustrations of this concern are:

> Both FED-STD-H28/20B [14] and ASME B1.3M-1992 [10] include the three screw-thread gaging systems: System 21; System 22; System 23. The type of guidance offered a user of each system in FED-STD-H28/20B, e.g., Section III-A of this report, would be available to a user of ASME B1.3M-1992 if there were a cross-reference in ASME B1.3M-1992 to FED-STD-H28/20B.
>
> Length-of-thread engagement during thread inspection is a critical factor influencing the outcome of thread inspection. However, the required length-of-thread-engagement during inspection is not explicitly stated in FED-STD-H28/20B or in ASME B1.3M-1992, nor is the source standardization document containing the specified length of thread engagement explicitly referred to in either standard.

This seems to be a matter for the custodians of standardization documents to address.

Issue V — Uncertainty about Error Propagation during Gage Calibration: A concern of long standing seems to be uncertainty about how error propagation in a sequence of calibrations, (i.e., calibration of master gages using gage blocks and calibration of working gages using master gages) affects reliability and repeatability of the calibration of the working gages.

ANSI/ASME B1.2-1983 [11] addresses calibration of thread-setting gages somewhat indirectly. For example, in B1.2-1983 [11], design of *W tolerance thread setting plug gages* is addressed in Section 5.13 on pages 143-148, and calibration of their elements is described in Table 14 on page 159. Design of W tolerance thread setting ring gages is not discussed in Section 4 of ANSI/ASME B1.2-1983, although the calibrations described in Table 14 could apply to W tolerance thread-setting ring gages.

ANSI/ASME B1.2-1983 [11] addresses calibration of *X tolerance working gages* in Table 12 (pages 143-154) for external product threads, and in Table 13 (pages 155-158) for internal product threads.

Further discussion of gage calibration appears in Appendix A on pages 163-171 of ANSI/ASME B1.2-1983 [11], but the concern about propagation of errors in a sequence of calibrations is not addressed.

---

[12] "System 21 provides for interchangeable assembly with functional size control at the maximum material limits within the length of standard gaging elements, and also control of the characteristics identified as **NOT GO** functional diameters."

This concern seems to be a matter to be addressed by: 1) the Quality Program in any institution which calibrates and maintains screw-thread gages and measuring equipment, and 2) the activity in ASME B1 which is documenting Gaging Uncertainty.

Issue VI — Shortage of Technical Data about Mechanical Performance: The shortage of public domain technical data which describes the mechanical performance of threaded assemblies is illustrated by the fact that the primary reference document about this concern seems to be a 1964 translation from Russian of the book by A.I. Yakushev, *Effect Of Manufacturing Technology And Basic Thread Parameters On The Strength Of Threaded Connections* [22].

As the present study progressed, it became evident that technical data about the mechanical performance of threaded assemblies exist in abundance. Such technical data have been generated for years and years. However, these technical data are regarded as highly proprietary by the originating institutions and remain under secure control of those institutions. This circumstance is unfortunate for the purposes of this study, for technical data in the form of failure analysis reports about incidents linked to alleged non-conforming threads are not available to be evaluated and compared with highly-visible and frequent public statements about alleged non-conforming threads. This is not to presuppose that such incidents do or do not actually exist.

Issue VII — Excessive Costs Meeting Some Requirements of MIL-SPECS: This concern was raised by the aerospace industry. Certain requirements in MIL-S-7742B and MIL-S-8879C [7,8] are said to require duplicate measuring of thread characteristics and to require actual measurements of dimensions or angles and recording thereof in lieu of determining if the thread characteristics are within tolerances.

This concern appears to be a matter to be addressed by the custodian of these military specifications, in cooperation with users of the specifications in the aerospace industry. In fact, some military specifications are currently being reviewed by custodians and users, who are exploring possible replacement of non-consensus specifications with consensus standards.

VI. CONCLUSIONS:

1. The demand for continued improvement in the performance of threaded products in aerospace applications during the last fifty years led to modifications in military specifications for threaded products. In 1973, screw thread gaging systems[13] were introduced for the first time into MIL-S-7742 and MIL-S-8879. In 1991, modifications were made to MIL-S-7742 and MIL-S-8879 which resulted in increased inspection severity. Limit inspections using limit gages have been the predominant screw-thread inspection procedure for more than one hundred years. A gradual shift from limit inspections using limit gages to variables inspections using indicating gages accompanied the introduction of screw-thread gaging systems.

2. Screw-thread gaging systems were first introduced in ASME *consensus standards* in 1979 (ASME B1.3/B1.3M), and in *non-consensus federal standards* (FED-STD-H28/20) in 1981. In both standards, the three gaging systems are termed: "System 21"; "System 22"; "System 23".

3. International competition with the domestic automotive industry led domestic producers in the 1980s to introduce statistical process control (SPC) in procurement procedures for commodities and products, including screw-thread products. SPC in the automotive industry served as a driving force for increased use of indicating thread gages, which were included in the screw-thread gaging systems of the consensus (ASME) and non-consensus (FED-STD-H28) screw thread standards.

4. The shift from limit inspections using limit gages to variables inspections using indicating gages requires ongoing adjustments via education throughout the domestic screw-thread user community, the metrology and gage-making community, and the screw-thread manufacturing industry. Seminars and courses given by disinterested parties are needed to provide information about screw thread standardization documents, especially the relatively new standardization documents specifying screw thread gaging systems. Improved understanding is needed in at least the following areas by operating personnel who are faced with vast quantities of threaded products every day: 1) how standardization documents are formally put into use, 2) design and construction of screw-threads, 3) goals and results of required inspection of thread-elements/characteristics by each gaging system, 4) design details of commonly-used gages and measuring equipment, 5) available choices of gages and measuring equipment, 6) the distinction between limit inspections and size inspections, and the choices available for each under Systems 21, 22, and 23, and 7) data on the affect on product performance of the screw-thread gaging system.

5. Two concerns expressed by metrologists should be addressed by the metrology/gaging-making community.

---

[13] The phrase, "screw thread gaging systems," was not actually used in military specifications. The terms used in 1973 for the inspection systems were: "Method A," "Method B," and "Method C."

a) Uncertainty about error propagation in a sequence of calibrations from gage blocks to setting masters, and from setting masters to working gages and to product threads.

b) Variations in results of thread inspections using indicating gages from different manufacturers.

6. The current climate of uncertainty and confusion in the national screw-thread community can be mitigated through improvements in the clarity of standardization documents (language/terminology) relating to screw-thread gaging systems. Changes in language and clearer definitions of terms would help to clarify what the standards aim to accomplish.

7. When properly used, current screw-thread gaging standards are satisfactory for their intended use.

VII. RECOMMENDATIONS

A. Near-Term Recommendations

1. One Consensus Standard for Screw-Thread Gaging Systems

A single consensus standard for screw-thread gaging systems would seem to be appropriate for the needs of the domestic screw-thread community. Consequently, it is recommended that FED-STD-H28/20B [14] and MIL-S-8879C [7] be phased out, and that ASME B1.3M-1992 [10] be retained as the national standard for screw-thread gaging systems.

To facilitate this transition, it is recommended that the benefits of having a single consensus standard for screw-thread gaging systems be provided to:

the policy-making levels of the custodians of standardization documents, i.e., ASME, FED-STD-H28, and the U.S. Air Force.

the documentary standards committees dealing with screw-thread standardization documents.

representatives of stakeholders by convening panels intended to encourage their support. Stakeholders would include all interested parties in the national and trans-national screw-thread community, i.e., custodians and caretakers, makers of gages and measuring equipment, manufacturers of screw-thread products, and users of screw-thread products.

2. Study of How Thread Tolerance Variations Affect Mechanical Performance of Threaded Assemblies

It is recommended that a panel be convened to develop strategies for acquiring existing technical data from institutions which generate and use such data every day. For example, data might be sought from agencies of the U.S. Government, especially military agencies, under Freedom of Information requests. Such data would be used for establishing an open data base operated by a disinterested, third-party manager designated by the national screw-thread community.

Data showing how variations in thread tolerances affect mechanical performance of threaded assemblies would be beneficial for the education of mechanical engineers, and for thorough and objective evaluation of public claims about shortcomings in mechanical performance of threaded assemblies.

3. Metrology Concerns

It is recommended that two concerns of metrologists, namely, error propagation during a sequence of gage calibrations and variations in results of thread inspections using indicating gages from different manufacturers, be addressed by an appropriate committee made up of metrologists. Perhaps the American Measuring Tool Manufacturers Association could provide oversight for this activity.

4. Technical Article about Standards for Gaging Systems

It is recommended that an article be written based on the findings of this study. The article would aim to help practicing engineers understand the goals and results of inspections by existing screw-thread gaging systems.

B. Far-Term Recommendations

1. Technical Data on How Thread Tolerances Affect Mechanical Performance

The near-absence of literature in the public domain addressing how variations in thread tolerances affect mechanical performance of threaded assemblies is surprising. Such literature would be beneficial for the education of mechanical engineers and for thorough and objective evaluation of public claims about shortcomings in mechanical performance of threaded assemblies.

Archival Literature: It is recommended that a research agenda be developed and carried out by a consortium of interested parties to develop technical data and an archival literature in the public domain which addresses how variations in tolerances of thread elements and thread characteristics affect mechanical performance of threaded assemblies.

Data Base: It is recommended that a public-domain data base be established which addresses how variations in tolerances of thread elements and thread characteristics affect mechanical performance of threaded assemblies. Such a data base would be operated by a disinterested, third-party manager established by the national screw thread community.

2. Experimental Studies of Threaded Product Failures It is recommended that a program be established by a consortium of interested parties to conduct experimental studies of threaded product failures and to establish a public-domain data base with the findings.

3. Gaging Systems of the Future It is recommended that a consortium of interested parties promote research intended to develop laser gaging systems and holographic gaging systems for screw-thread inspection. An in-progress program for development of a laser gaging system for external product thread [34] could serve as a model for further research.

REFERENCES

1. ANSI/ASME B1.7M-1984 — *NOMENCLATURE, DEFINITIONS, AND LETTER SYMBOLS FOR SCREW-THREADS*. American Society of Mechanical Engineers, 345 East 47th Street, New York, New York 10017.

2. NATIONAL BUREAU OF STANDARDS HANDBOOK H25 — *SCREW-THREAD STANDARDS FOR FEDERAL SERVICES-1939*. National Institute of Standards and Technology, Gaithersburg, Maryland, 20899-0001.

3. ASME B1.1-1989 — *Unified Inch Screw-Threads (UN and UNR Thread Form)*. American Society of Mechanical Engineers, 345 East 47th Street, New York, New York 10017.

4. DuPont, B.R., Private communication, 1995.

5a. NATIONAL BUREAU OF STANDARDS HANDBOOK H28 — *SCREW-THREAD STANDARDS FOR FEDERAL SERVICES-1942*. National Institute of Standards and Technology, Gaithersburg, Maryland, 20899-0001.

5b. NATIONAL BUREAU OF STANDARDS HANDBOOK H28 — *SCREW-THREAD STANDARDS FOR FEDERAL SERVICES* — PART I-1957, National Institute of Standards and Technology, Gaithersburg, Maryland, 20899-0001.

5c. NATIONAL BUREAU OF STANDARDS HANDBOOK H28 — *SCREW-THREAD STANDARDS FOR FEDERAL SERVICES* — PARTS I-II-III-1969. National Institute of Standards and Technology, Gaithersburg, Maryland, 20899-0001.

6. FED-STD H28 — *SCREW-THREAD STANDARDS FOR FEDERAL SERVICES*. (Currently consisting of 23 sections.) Defense Industrial Supply Center, 700 Robbins Avenue, Philadelphia, PA 19111-5096.

7. MIL-S-8879C 25 JULY 1991: MILITARY SPECIFICATION — *SCREW-THREADS, CONTROLLED RADIUS ROOT WITH INCREASED MINOR DIAMETER, GENERAL SPECIFICATION FOR*. Department Of The Air Force, Headquarters United States Air Force, Washington, D.C., 20330.

8. MIL-S-7742D 25 JULY 1991: MILITARY SPECIFICATION — *SCREW-THREADS, STANDARD, OPTIMUM SELECTED SERIES: GENERAL SPECIFICATION FOR*. Department Of The Air Force, Headquarters United States Air Force, Washington, D.C., 20330.

9. International Standard ISO 1502 — *ISO General Purpose Metric Screw-Threads Gauging*. International Standards Organization, Case Postal 56, Geneve 20, CH-1211, Switzerland. (15 July 1978).

10. ASME B1.3M-1992 — *Screw-Thread Gaging Systems for Dimensional Acceptability — Inch and Metric Screw-Threads (UN, UNR, UNJ, M, and MJ).* American Society of Mechanical Engineers, 345 East 47th Street, New York, New York 10017.

11. ANSI/ASME B1.2-1983 — *Gages and Gaging for Unified Inch Screw-Threads.* American Society of Mechanical Engineers, 345 East 47th Street, New York, New York 10017.

12. ASME B1.18M-1982 — *Metric Screw-Threads for Commercial Mechanical Fasteners - Boundary Profile Defined* (withdrawn). American Society of Mechanical Engineers, 345 East 47th Street, New York, New York 10017.

13. ASME B1.19M-84 — *Gages for Metric Screw-Threads for Commercial Mechanical Fasteners - Boundary Profile Defined* (withdrawn). American Society of Mechanical Engineers, 345 East 47th Street, New York, New York 10017

14. FED-STD-H28/20B 10 MARCH 1994 — SCREW-THREAD STANDARDS FOR FEDERAL SERVICES - SECTION 20 - *INSPECTION METHODS FOR ACCEPTABILITY OF UN, UNR, UNJ, M, AND MJ SCREW-THREADS.* Defense Industrial Supply Center, 700 Robbins Avenue, Philadelphia, PA 19111-5096.

15. ASME B1.16-1984 — *Gages and Gaging for Metric M Screw-Threads.* American Society of Mechanical Engineers, 345 East 47th Street, New York, New York 10017.

16. ASME B1.22M-85 — *Gages and Gaging for Metric MJ Series Metric Screw-Threads.* American Society of Mechanical Engineers, 345 East 47th Street, New York, New York 10017.

17. "Recommendations for Fastener Thread Acceptability." Manual compiled and published by the Industrial Fasteners Institute, 1505 East Ohio Building, 1717 East Ninth Street, Cleveland, Ohio 44114, 1990.

18. *Review Of MIL-S-8879C And MIL-S-7742D.* Aerospace Industries Association, I Street, Washington, DC, March 16, 1993.

19. Ellison, H.W., *The Effect Of Lead And Pitch Diameter On Bolt And Nut Strength.* Internal Report, General Motors Corporation, December 15, 1969.

20. FED-STD-H28/2B, SCREW-THREAD STANDARDS FOR FEDERAL SERVICES, SECTION 2, *UNIFIED INCH SCREW-THREADS,* Defense Industrial Supply Center, 700 Robbins Avenue, Philadelphia, PA 19111-5096.

21. Bickford, J.H., *An Introduction to the Design and Behavior of Bolted Joints,* Marcel Dekker, Inc., New York, 1990.

22. A.I. Yakushev, *Effect of Manufacturing Technology and Basic Thread Parameters on the Strength of Threaded Connections,* The Macmillan Company, New York, 1964.

23. French, M., *Form, Structure, and Mechanism*, Chapter 3, Section 3.3, pp. 56-58, "Screwed Fastenings," Springer-Verlag, New York, 1992.

24. Cullum, R.D., *Handbook of Engineering Design*, Chapter 5, Section 5.1, "A Logical Approach To Secure Bolting," pp. 107-108, Butterworth & Co. (Publishers) Ltd., 1988.

25. Kingsbury, N.R., *MILITARY FASTENERS — Changes to Specifications Are Justified*, Report from the United States General Accounting Office to the Chairman, Subcommittee on Investigations, Committee on Armed Services, House of Representatives, September, 1991.

26. Private communication, *A White Paper, Fastener Strength Analysis*, Nuclear Safety Concern 93-11, U.S. Nuclear Regulatory Commission, Washington, D.C. 20555-0011, June 9, 1994.

27. Private communication, *Report To ASME Special Committee To Study Development Of An Optimum Metric Fastener System On Size Of Production Screw-Threads*, August 16, 1972.

28a. Alexander, E.M., *Design and Strength of Screw-Threads*, Transaction of Technical Conference on Metric Mechanical Fasteners, ANMC, 17-19 March 1975.

28b. Alexander, E.M., *Analysis and Design of Threaded Assemblies*, Paper No. 770420, Society of Automotive Engineers, Inc., 400 Commonwealth Drive, Warrendale, PA, 15096, 1977.

29. Greenslade, J., and Wilson, C.J., Private Communications, October 1994, and December 1995.

30. Johnson, S., Private Communication, November 1994.

31. Matievich, W., Private Communication, *Test Data of Internal Threads*, January 1993.

32a. Letter from Mr. John Simpson, Acting Chief, Mechanics Division, National Bureau of Standards, Gaithersburg, Maryland to Mr. C.T. Gustafson, Quality Assurance Office, Portsmouth Naval Shipyard, Portsmouth, New Hampshire, June 27, 1977.

32b. Letter from Mr. John Simpson, Director, Manufacturing Engineering Laboratory, National Institute of Standards and Technology, Gaithersburg, Maryland, to the Honorable Donald B. Rice, Secretary of the Air Force, The Pentagon, Washington, D.C., July 1, 1991.

32c. Letter from Mr. Richard H.F. Jackson, Deputy Director, Manufacturing Engineering Laboratory, National Institute of Standards and Technology, Gaithersburg, Maryland, to Mr. James A. Davis, Office of Nuclear Reactor Regulation, Nuclear Regulatory Commission, Washington, D.C., March 10, 1994.

33. Veale, R., Strang, A., and Hsiao-Yu, C., Private Communication, *Variations In Size Measurements By Indicating Gaging Systems*, National Institute of Standards and Technology, Gaithersburg, Maryland, 20899-0001. August 1994.

34. Castore, G., Private Communication, January 1995.

APPENDIX A

INSPECTABILITY OF SCREW THREAD ELEMENTS AND CHARACTERISTICS USING GAGES AND MEASURING EQUIPMENT SPECIFIED IN CURRENT DOMESTIC STANDARDS FOR SCREW THREAD GAGING SYSTEMS

CONCERNS: The following concerns about the capabilities of screw thread gages and measuring equipment specified in current domestic standards for screw thread gaging systems have been expressed by an expert in screw thread inspection technology.

"Screw threads are geometrically complex and gages used to check them are varied in design and function. They all use comparison to setting or master gages which interact with the working gages. Pitch diameters of the setting gages cannot be "perfect" and the working gage designs may not even check pitch diameter as a single element. There is much variation in gage design and there are many variables that interact between setting gages, working gages and product gages. System 22 does not assure anything nor does System 23. The only thing definite about the three systems is that more elements are checked with System 23 than with System 22 and more with System 22 than with System 21."

SCREW THREAD GAGES AND MEASURING EQUIPMENT: Current domestic standards for screw thread gaging systems can be found listed in Tables II-2A and II-2B of this report. Three screw thread systems are currently in use: Systems 21, 22, and 23.

Screw thread elements and screw thread characteristics which might be specified for inspection appear in each of these standards. Tabulations of thread elements/characteristics cited in ASME B1.3M-1994 for external product threads appear in Tables III-1A and III-1B of this report.

Tables III-1A and III-1B also list the reference number of the screw thread gages and measuring equipment specified for inspecting thread elements/characteristics of external product threads in ASME B1.3M-1994.

Essential specifications and dimensions for the screw thread gages and measuring equipment specified in ASME B1.3M-1994 can be found in ANSI/ASME B1.2-1983, *GAGES AND GAGING FOR UNIFIED INCH SCREW THREADS.*

> The types of thread gages and measuring equipment indicated by the reference numbers can be determined from Tables 1 and 2 in ANSI/ASME B1.2-1983.
>
> Tables 12 and 13 (pages 151-154 and 155-158, respectively) of ANSI/ASME B1.2-1983 list calibration requirements for the thread gages and measuring equipment.

PROPOSED STUDY:

The proposed study would measure the effectiveness of the various gages for thread tolerance zone evaluation.

> A) The design and measuring capability of each gage and each kind of measuring equipment used to inspect screw threads, as listed in Tables III-1A and III-1B of this report, and as listed in Tables 1, 2, 3, and 4 of ASME B1,3M-1994, be evaluated by at least three disinterested metrology experts who then prepare a report summarizing their findings which is issued in the public domain.
>
> B) A data base be established in the public domain which addresses the designs and measuring capabilities of the gages and measuring equipment listed in Tables 1, 2, 3, and 4 of ASME B1.3M-1994.

APPENDIX B - GLOSSARY[14]

ACCEPTABILITY
ACCEPTABLE GO FUNCTIONAL LIMIT OF THREAD
ACCEPTABLE GO FUNCTIONAL SIZE OF THREAD
ACCEPTABLE NOT GO FUNCTIONAL LIMIT OF THREAD
ACCEPTABLE NOT GO FUNCTIONAL SIZE OF THREAD

ATTRIBUTES OF A THREADED PRODUCT
ATTRIBUTES INSPECTION
ATTRIBUTES/FIXED LIMIT CONTROL
CIRCULARITY (ROUNDNESS)
CUMULATIVE FORM VARIATION
CUMULATIVE ELEMENTS VARIATION

DIMENSIONAL CONFORMANCE

ENVELOPE
SCREW-THREAD ENVELOPE
THREAD DEFINITION ENVELOPE

FLANK ENGAGEMENT

FUNCTIONAL DIAMETER

ACTUAL PRODUCT SCREW-THREAD (IMPERFECT SCREW-THREAD)

LENGTH OF GAGE

LIMITS OF SIZE

LIMIT INSPECTION

MAXIMUM MATERIAL

MINIMUM MATERIAL

PITCH DIAMETER

IDEAL SCREW-THREAD (PERFECT SCREW-THREAD)

SINGLE ELEMENT GAGING (7742B-1968/¶3.5.2)
SIZE INSPECTION

STRAIGHTNESS

SCREW-THREAD ELEMENTS
SCREW-THREAD CHARACTERISTICS

THREAD DEFINITION ENVELOPE

THREAD INDICATING GAGES
THREAD LIMIT GAGES

TOLERANCE ZONE
VARIABLES OF A THREADED PRODUCT
VARIABLES INSPECTION
VARIABLES/INDICATING CONTROL

---

[14] Definitions on the following pages are grouped by relationship rather than in alphabetical order.

ACCEPTABILITY — an indication that a threaded product has passed the inspection requirements specified in the selected gaging system in the designated inspection standard.

DIMENSIONAL CONFORMANCE — an indication that thread elements and thread characteristics of a threaded product have sizes within the maximum and minimum limits of size which are specified in the applicable thread standard.

ENVELOPE — a solid helix in space. The inner and outer radial boundaries of the helix are given by perfect thread profiles at the minimum and maximum pitch diameter, respectively, of a screw-thread. The longitudinal axis of this helix coincides with the longitudinal axis of the screw-thread, and the pitch of this helix is identical to the pitch of the screw-thread. The envelope extends longitudinally over the entire product thread length.

LIMITS OF SIZE — the envelope defines the limits of size of a screw-thread.

SCREW-THREAD ENVELOPE — same meaning as "envelope".

TOLERANCE ZONE — same meaning as "envelope"

SIZE — a designation of magnitude.

SCREW-THREAD ELEMENTS — pitch diameter, pitch, flank angle, lead angle, major diameter, minor diameter, root, crest.

SCREW-THREAD CHARACTERISTICS — functional diameter; axial taper; roundness (circularity); straightness; surface finish; nicks, breaks and burrs.

PITCH DIAMETER - an *element* and a *feature* of an actual product screw-thread; the diameter of an imaginary cylinder, the surface of which would pass through the threads at such points as to make equal the width of the thread ridges and the width of the gaps (grooves) between thread ridges.[15]

PITCH CYLINDER — a cylinder with diameter equal to the pitch diameter. The *pitch cylinder* is a fundamental construction feature of a screw-thread.

FUNCTIONAL DIAMETER - the pitch diameter for the smallest internal perfect thread form which can engage the external product thread or the largest external perfect thread form which can engage the internal product thread. This size will be controlled by the amount of thread form imperfections in the product thread and is an indication of the fit with the mating thread. Functional diameter is termed the *thread assembly diameter* in ANSI/B1.7M-1984.

THREAD ELEMENTS — pitch diameter, pitch, flank angle, lead angle, major diameter, minor diameter, root, crest, surface finish.

CUMULATIVE ELEMENTS VARIATION — the combined effect on design pitch diameter of a screw thread accruing from individual variations in lead (including helix variation) and flank angle. This combined effect contributes to the magnitude of the thread characteristic termed *functional diameter*.

---

[15]This definition and discussion of it appears in the book by Werner F. Vogel, The Exact Over-Wire Measurement of Screws, Gears, Splines, and Worms, (Wayne State University Press, Detroit) 1973. See page 26.

CUMULATIVE CHARACTERISTICS VARIATION — the combined effect on pitch diameter of a screw-thread accruing from individual variations in axial taper, roundness (circularity), deviations from straightness, and from nicks, breaks, and burrs. This combined effect contributes to the magnitude of the thread characteristic termed ***functional diameter***.

IDEAL SCREW-THREAD — a thread with no cumulative variations, i.e., no variations in thread elements or in thread characteristics. The *functional diameter* of an ideal screw-thread is equal to the pitch diameter of the ideal screw thread. (The ideal screw-thread exists only in imagination.)

PERFECT SCREW-THREAD — same as *ideal screw-thread* but fitting within the thread tolerance zone.

LIMIT INSPECTION — qualitative examination of a threaded product with *thread limit gages* or with *thread indicating gages* to determine if the thread is inside or outside the inner and outer radial boundaries of the envelope.

SIZE INSPECTION — quantitative examination of a thread product to measure the sizes of specific thread elements or thread characteristics. Results of size inspection can be used for determining dimensional conformance with limits of size specified in the applicable thread standard.

LENGTH OF GAGE — the active linear distance from one end to the other of a *thread limit gage* or *thread indicating gage* (ring, plug, roll, or segment) which is used for thread inspection. The length of a gage determines the number of threads engaged during inspection.

ATTRIBUTES OF A THREADED PRODUCT — thread characteristics or thread elements.

ATTRIBUTES INSPECTION — an inspection method for attributes of a threaded product using specified thread limit gages or measuring equipment including indicating gages to determine if thread elements or thread characteristics are within specified limits of size.

VARIABLES OF A THREADED PRODUCT — thread elements or thread characteristics.

VARIABLES INSPECTION — an inspection method for variables of a threaded product using specified thread measuring and gaging equipment to establish the sizes of thread elements or thread characteristics.

LIST OF PUBLICATIONS
ASME CENTER FOR RESEARCH AND TECHNOLOGY DEVELOPMENT

Volume No.	Title
CRTD-1	*Consensus on Current Practices for Lay Up of Industrial and Utility Boilers,* 1985, 28 pp. Book Number H00336
CRTD-2	*Research Needs in Thermal Systems,* 1986, 221 pp. Book Number I00212
CRTD-3	*Goals and Priorities for Research in Engineering Design,* 1986, 164 pp. Book Number I00230
CRTD-4	*Pressure Systems Energy Release Protection (Gas Pressurized Systems),* 1986, 425 pp. NASA Contractor Report 178090
CRTD-5	*Thermodynamic Data for Biomass Materials and Waste Components,* 1987, 496 pp. Book Number H00409
CRTD-6	*Analysis of the Feasibility of Replacing Asbestos in Automobile and Truck Brakes,* 1988, 136 pp. Book Number I00264
CRTD-7	*Hazardous Waste Incineration: A Resource Document,* 1988, 192 pp. Book Number I00266
CRTD-8	*Research Needs in Dynamic Systems and Control – Volume 1: Strategic Initiatives and Opportunities,* 1988, 58 pp. Book Number I00275
CRTD-9	*Research Needs in Dynamic Systems and Control – Volume 2: Acoustics and Noise Control,* 1988, 48 pp. Book Number I00276
CRTD-10	*Research Needs in Dynamic Systems and Control – Volume 3: Control of Mechanical Systems,* 1988, 40 pp. Book Number I00277
CRTD-11	*Research Needs in Dynamic Systems and Control – Volume 4: Machine Dynamics,* 1988, 37 pp. Book Number I00278
CRTD-12	*Research Needs in Dynamic Systems and Control – Volume 5: Nonlinear Dynamics,* 1988, 32 pp. Book Number I00279
CRTD-13	*Peer Review Thermal Energy Storage Program,* 1989, 46 pp. Available from CRTD, Washington, D.C.
CRTD-14	*The ASME Handbook on Water Technology for Thermal Power Systems,* 1989, 1900 pp. Book Number I00284
CRTD-15	*Research Needs and Technological Opportunities in Mechanical Tolerancing,* 1990, 51 pp. Book Number I00298
CRTD-15-1	*Selected Case Studies in the Use of Tolerance and Deviation Information During Design of Representative Industrial Products,* 1992, 156 pp. Book Number I00329
CRTD-15-2	*Mathematization of Dimensioning and Tolerancing,* 1992, 38 pp Book Number I00337
CRTD-16	*Engineering Fluid Mechanics Workshop Report,* 1990, 114 pp. Book Number I00299
CRTD-17	*ASME Research Committee on Corrosion and Deposits From Combustion Gases Seminar of Fireside Fouling Problems,* 1990, 183 pp.
CRTD-18	*ASME Ash Fusion Research Project,* 1990, 11 pp. Book Number I00307
CRTD-19	*Achievements in Tribology,* 1990, 173 pp. Book Number H00269
CRTD-20-1	*Risk-Based Inspection – Development of Guidelines: Volume 1, General Document,* 1991, 155 pp. Book Number I00311
CRTD-20-2	*Risk-Based Inspection – Development of Guidelines: Volume 2, Part 1, Light Water Reactor (LWR) Nuclear Power Plant Components,* 1992, 156 pp. Book Number I00321
CRTD-21	*Hazardous Waste Destruction by High Temperature Incineration,* 1991, 21 pp. Book Number I0266A

LIST OF PUBLICATIONS
ASME CENTER FOR RESEARCH AND TECHNOLOGY DEVELOPMENT

Volume No.	Title
CRTD-22	*Mining Workshop for Nuclear Waste Cleanup,* 1991, 145 pp. Book Number I00314
CRTD-23	*The Use of Decision-Analytic Reliability Methods in ASME Codes and Standards Work,* 1993, 120 pp. Book Number I00351
CRTD-24	*ASME/Bureau of Mines Investigative Program on Vitrification of Residue From Municipal Waste Combustion Systems,* 1994, 132 pp. Book Number I00373
CRTD-25	*Installation of Plastic Gas Pipeline in Steel Conduits Across Bridges,* 1993, 64 pp. Book Number I00353
CRTD-26	*Lubrication Technology for Advanced Engines – An Assessment of Industrial Needs,* 1993, 104 pp. Book Number I00352
CRTD-27	*1993 International Forum on Dimensional Tolerancing and Metrology,* 1993, 324,pp. Book Number I00360
CRTD-28	*Research Needs and Opportunities in Friction,* 1993, 48, pp Book Number I00365
CRTD-29	*Research Guidelines for Aluminum Product Applications in Transportation and Industry,* 1994, 240 pp. Book Number I00366
CRTD-30	*Tribology in Manufacturing Processes,* 1994 Book Number I00374
CRTD-31	*An Evaluation of the Cost of Incinerating Wastes Containing PVC,* 1994 Book Number I00377
CRTD-32	*Hazardous Waste Incineration: What Engineering Experts Say,* 1994, 17pp. Book Number I0266B
CRTD-33	*Final Report From the Aluminum Industry Workshop,* 1994, 273 pp.
CRTD-34	*Consensus on Operating Practices for the Control of Feedwater and Boiler Water Chemistry in Modern Industrial Boilers,* 1994, 38 pp. Book Number I00367
CRTD-35	*A Practical Guide to Avoiding Steam Purity Problems in the Industrial Plant,* 1995, 40 pp. Book Number I00383
CRTD-36	*The Relationship Between Chlorine in Waste Streams and Dioxin Emissions From Waste Combustor Stacks,* 1995, 716 pp. Book Number I00385
CRTD-37	*Screw-Thread Gaging Systems for Determining Conformance to Thread Standards,* 1996, 72 pp. Book Number I00391